W9-BUZ-764

# GLORIOUS BEEF

## THE LAFRIEDA FAMILY AND THE EVOLUTION OF THE AMERICAN MEAT INDUSTRY

*Pat LaFrieda*

WITH CECILIA MOLINARI

*An Imprint of* HarperCollins*Publishers*

FIRST EDITION

*Designed by Michelle Crowe*

*Illustrations by IvanDbajo, jenesesimre / Adobe Stock*

*Background texture by schab / Adobe Stock*

Library of Congress Cataloging-in-Publication Data has been applied for.

ISBN 978-0-06-296670-4

21 22 23 24 25 LSC 10 9 8 7 6 5 4 3 2 1

TO MY DAD,

FOR MAKING THIS ALL POSSIBLE

# CONTENTS

PART THREE:

## TO YOUR TABLE



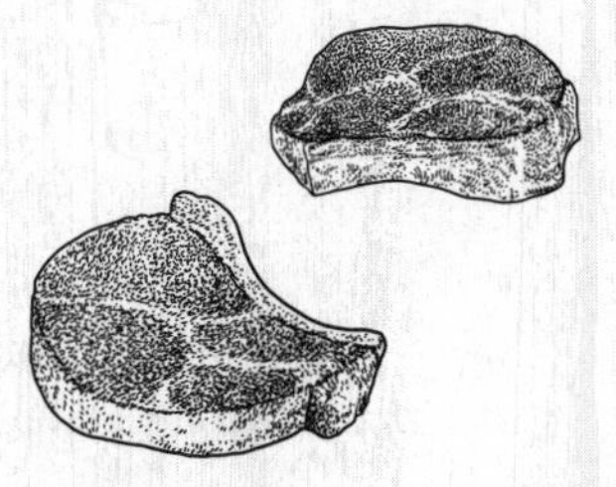

# INTRODUCTION

It all began when my great-grandfather Anthony LaFrieda decided to pack up and move his family from Italy to New York in 1909 in search of a better life. He set up our family's first retail butcher shop in 1922 in Flatbush, Brooklyn. Almost one hundred years later, here I am, a fourth-generation butcher, third-generation meat purveyor, at the helm of a family-run business that has been providing meat to customers in America for decades. My family has worked through wars, the Great Depression, the tumultuous years when New York City was dubbed Fear City, the fall of the Twin Towers, unprecedented hurricanes, and even a pandemic, and we're still going strong.

During the late 1990s, when rents in the Meatpacking District skyrocketed and small businesses were being taken over by large impersonal corporations, Pat LaFrieda Meat Purveyors remained in the family and went on to become enormously successful. When my father started working with my grandfather, they were located in a small spot inside a decrepit old building amid the chaos of the Fourteenth Street Meat Market. My dad went from working in extreme conditions with no heat or elevator to a 35,000-square-foot facility in New Jersey, right across the river from Midtown Manhattan, and we've now opened a second facility two blocks away. With close to two hundred employees, we provide more than five hundred thousand pounds of meat a week to retailers, restaurants, and home delivery customers. We are a bridge between meat and people.

Our national meat industry is an intricate yet remarkably efficient system that feeds millions of people across the country. When I joined the business in 1994, no one really cared about where the meat came from—they just cared about the quality. But the quality is directly linked to how those animals were raised, finished, and harvested.

I first set out to share what I had been learning on the field with my clients many years ago. The more they knew, the better choices they would make for their businesses and patrons. In order to keep them informed on the latest trends and innovations and the industry's ever-evolving new technology and advances, I need to constantly stay up to date on what is happening in our industry as a whole.

Beyond that, my meat purveyor role often extends into consulting, with chefs reaching out to brainstorm what they could put on the menu of their new restaurants or how they could revamp their old menus to attract more customers. That's how many famous dishes came to be, including the Black Label Burger at Minetta Tavern, the Shake Shack all-natural burger, and our signature Original Filet Mignon Steak Sandwich.

Most people don't know the amount of time, commitment, and work that goes into the piece of meat on their plate. I'm here to explain it to you. This isn't your average farm-to-table book. It's not for the faint of heart. I want to take you behind the scenes of each stage of the process: the ups, the downs, the struggles, the accomplishments, and the misconceptions surrounding beef related to everything from our health and the environment to what you're really buying at the store.

What are the real implications of grass-fed beef on climate change? What goes into humanely processing animals at harvesting facilities? Why are grading, labeling, and traceability essential for the end consumer? *What's the beef with eating meat?* There are two sides to every story, yet, in the beef industry's case, only one seems to get most of the airtime. I want to help correct this mistake and reveal the side that is usually ignored by the media, which often presents a simplistic, one-dimensional story.

That's why *Glorious Beef* is not just about debunking myths; it's about sharing the truths behind the industry's story of survival and constant evolution. It's based on what I have seen and experienced on a personal level as I grew up and on a professional level as I came up in the industry myself. I'm not here to tell you what to do or how to think—I want to give you enough information for you to be able to make the right choices for yourself. It's time to bring beef back to its glory, celebrate its history and accomplishments, and shine a light on all it has to offer. This is the industry's story. This is the story of everyone who enjoys that juicy steak on their plate. And, of course, this is also my story, because the meat industry is my home.

PART ONE
FROM THE FARM

# 1

# THE GROWERS

It's three in the morning and I'm sitting in the passenger's seat of my dad's car, heart pounding with excitement as he drives down the Belt Parkway from our home in Bensonhurst, Brooklyn, toward Manhattan. I grab my dad's pack of Marlboro Reds lying on the seat between us and take a long and deep whiff. I've never smoked, but that smell of raw tobacco to this day is my dad, even though he quit years ago; it bathes me in a sense of pure comfort. Wide-awake with anticipation, I look out at the empty New York streets in the dark of the night. I've driven to work with my dad before, but this time it's different. It's 1981, I'm ten years old, and my father is going to finally let me stand at the butcher's table with him and my grandfather and cut meat for our customers.

We walk into our family shop, Pat LaFrieda Meat Purveyors, on Bleecker Street and West Tenth and, as my dad greets my grandfather and settles in, I go straight for the butcher coats, sliding my arms into one of those all-important white garments and buttoning it up over the several layers of T-shirts and shirts I had carefully put on earlier. I'd learned my lesson: if I wanted to make it through the work shift, I knew I had to layer up and keep moving to stay warm in the 36-degree temperatures.

When my dad gives me the signal, I walk over to the six-foot-long butcher table lined with meat that is ready to be turned into the day's orders. At ten, I barely reached the table, but I got the boost I needed by stepping onto an old milk crate that propelled me up to the cutting sweet spot where the knives were at a safe distance from my face. And there I stood, flanked by my father at right and my grandfather at left—the two men I admired most in the world—immersed in the smell of sweet fresh beef. As I sank my knife into the top round before me, all I could think was, *This is it. This is what I want to do with my life.*

I was hooked.

But I knew nothing about where the meat on the table came from; I had no idea about all the hard work and sacrifice that went into procuring it, selling it, and delivering it. For a born and raised Brooklyn kid, it was hard to imagine farms filled with herds of cattle, the far-reaching paddocks out west dotted with cows and calves, bullocks and bulls, steers and heifers. I probably didn't even know the terminology back then, but I have come to believe that if you really want to understand something, you have to go to the source, and that source, in our case, is the growers.

I'm often asked, "Pat, where do you get your cows from?" Generally speaking, we do not eat cows. Stay with me. A cow is a female that has had at least one calf. A bull is a mature male used for breeding. These two animals mate and produce calves, which grow into steers and heifers. A steer is a castrated male (prior to castration they are called bullocks) and a heifer is a female that hasn't had a calf yet. So, if you happen to have a piece of high-quality meat on your plate, then you are most certainly eating a steer or heifer, not a cow. Cows will show up in meat products, but mainly lower-quality ones; they don't make the cut when it comes to Prime beef in large part because they're too old at the time of harvesting.

And who are the growers? The growers are the farmers, the owners of herds of cattle that feed the more than 330 million people in our country. How do they do this? First, they have to identify what type of product they want to sell and where it will go.

In cattle-speak, the two basic categories we deal with in the industry are beef cattle and milk cattle, named for what they produce. Next, they have to figure out what breed of cattle they want to raise. To keep it simple, let's focus on the basic three that we deal with in our business: Holstein, Hereford, and black Aberdeen Angus. Holstein are your classic milk cattle; this is also the breed used for veal. Hereford and black Aberdeen Angus are beef cattle.

I prefer black Aberdeen Angus, or Black Angus for short, for LaFrieda Meats. Hereford is a solid breed, but Black Angus has the best yield (meaning it has the best usable meat-to-waste ratio) and amazing marbling, which is the streaks of intramuscular fat you can see in a piece of beef. Reaching the decision to use Black Angus took time and experience, most of which I amassed from the training I got back in the 1980s and 1990s when LaFrieda Meats was located in the West Village in New York and we bought our product directly from the infamous Fourteenth Street Meat Market, in what's now known as the Meatpacking District. Long gone are the days of carcasses on hooks lining those streets, but I remember it like it was yesterday: it was grimy and gritty, and it was all we had when it came to procuring our meat.

After a few years on the job picking up beef at the Meat Market, I began to notice that the Black Angus breed always had a stern, large eye—aka the longissimus muscle, which extends from the front shoulder to the hind leg and appears as an eye of meat between the forequarter and hindquarter—and it continued all the way down into the loin, reaching the porterhouse. This breed also always rendered larger filets than the Herefords. For that reason alone, I started to request more Black Angus beef at the market. That's when we started carrying larger filets and larger porterhouse steaks, and we even got some chuck steaks that were almost as big as rib eyes at a quarter of the price. Since Herefords have a higher spine—or what we call a razor back, which is when the hip bones poke out from the animal's rear end—their eye tapers off and isn't as full as the one in a Black Angus. Bonier cattle means they yield less meat. Yes, the Black Angus cost more, but its yield is so impressive that buying more of that breed just made basic sense. And that became one of our first important specifications in the business: procuring Black Angus with black hides. The reason we request black hides is that this helps distinguish them at the harvesting facility from other cattle, such as Hereford, which have red hides.

Around this time, I also began to realize that the way to have the most control over our product was to once and for all wean ourselves off the Fourteenth Street Meat Market and go straight to the source, to the growers. It all began with veal.

Twenty years ago, we specialized in lamb and veal, which we were buying from beef handlers at the Meat Market. Veal was delivered to three distributors in Manhattan—this meant that I had to hit all three to buy our shop's supply of veal and hope I was lucky enough to get there before the other 250 meat purveyors in New York City tagged the best pieces of meat for themselves. There were no schedules and we never knew when they got their deliveries, so it wasn't as easy as just showing up at a specific time. Plus, I couldn't be at three places at once tagging meat, so I had to choose one knowing I would have to settle with what was left at the other distributors when I finally got around to them. As if this wasn't enough, the pecking order would come into play. The distributors always set aside the best veal for their cream customers, the ones who had good credit and bought more than the rest, so the small guy usually got the short end of the stick. Back then, that small guy was us, which made getting consistently good veal difficult. Even worse, some of these distributors had the foresight to own and operate separate companies under different names, which sold directly to restaurants, so suddenly we were competing with the same people who were selling us our meat. We knew what they were doing—everyone did—but we had no other choice.

After riding out what felt like hours of traffic on those Manhattan streets and finally picking out and tagging my veal of choice, I'd come by later to pick it up only to find they'd switched the piece for another one while packaging it. This happened all the time, and those who quietly had to pay the price were us meat purveyors.

New York is the best place to learn the ropes of the beef business because there's no messing around and no exceptions: you have to carry the best or you're out. That's what happened at one of the city's top-rated restaurants, Il Mulino on West Third Street. They were renowned in the industry for produce, seafood, and meat, especially their huge domestic lamb chops and veal chops, and they only accepted the best. If they got something they didn't like, they just handed it right back to you. They didn't want to hear any excuses, they didn't want to know about the Meat Market drama and corruption—you either replaced the product or they walked. That's how my dad got them as a client to begin with. On one occasion, their supplier didn't deliver the right veal and they ended that long-term relationship without a flinch. Compromising their restaurant's quality was out of the question.

Whoever supplied Il Mulino back then was seen as a great meat purveyor because those guys were tough. They lived above the restaurant, worked fifteen- to sixteen-hour days, and were always on top of their game. So, once my dad landed them as clients, we knew we had to do everything in our power to keep them happy because having them in our roster automatically spoke highly about us. That is what ultimately drove me to think beyond the godforsaken Meat Market.

I knew we needed our veal to be as pale as possible, pinkish as opposed to reddish. Veal's color is connected to the amount of iron in the muscle; the more iron, the redder the veal, and by Italian American standards red veal is perceived as lower quality. I also knew the veal needed to be milk fed and around ten months of age. My dad thought I was crazy to think we could pursue growers directly, but he supported my nutty idea and came with me to Pennsylvania to find the veal that would meet our standards and keep Il Mulino happy.

We drove to Lancaster, a place I knew well because I had been stationed there in the army. I joined the army reserve as a combat medic when I was nineteen, served for nine years, and was honorably discharged in 1999 when I decided to go full steam ahead with the family business. On my weekend drills as a sergeant, there was a guy in the motor pool who used to get me out of the base by giving me a five-ton truck and a list of small tasks, like picking up a water chest from another base. Three or four army personnel were required to move this type of truck, one stationed in the back to direct me and the others to operate different parts of the vehicle. So, I'd pick the soldiers and we'd all climb in and head out for a ride around Pennsylvania. That's how I got to know Lancaster. As I drove around, I began to notice the different farms in the area with cages in the middle of their fields and figured out that was for the veal. So, when this idea of going straight to the source hit me, I knew exactly where to head to first.

My dad wasn't all that pleased when I joined the army; the thought that I could be deployed and something could happen to me worried him. So that trip was also the first time I was able to take my dad to visit my base and show him what I'd been doing all those years. Eventually, we found some Amish growers and dropped in on them without warning. At first, they looked at us like we had three heads. They had never encountered a meat purveyor who sold to restaurants at our level—which back then was small—and went to their farms in person to ask them questions about their veal in search of a specific size in the veal chop's eye diameter. We were basically sharing with them the exact specifications of what we needed, a practice that wasn't at all common at the time.

Through the course of this conversation, we realized that they were under the impression that purveyors like us wanted lighter veal (this idea may have come from French restaurants requesting smaller veal chops). When we told them that we actually wanted heavier ones, they were slightly taken aback, but they were willing to work with us to make it happen. Now that we had started the conversation with the growers, we needed to figure out how to get the veal harvested under USDA guidelines, stamped, and delivered to us in Manhattan. I'll get into more detail regarding this process in the following chapters, but suffice to say, in order for this plan to succeed, we had to find a slaughter facility that worked with the growers we had spoken to. That's how we first met Tony Catelli, the owner of this establishment. He must've thought there was something wrong with us. Why weren't we buying straight from the Fourteenth Street Meat Market? Did we not have the money to back up our requests? But we immediately put his mind at ease by telling him that we'd pay in advance. We gave him our specs so that he could measure the diameter of the eye and pick the veal that would best serve us and in turn fronted him $20,000 to prove we were legitimate.

That's how we first broke free from the Fourteenth Street Meat Market trap. Relying on only three companies to determine my future? That wasn't going to happen. Going straight to the growers for sourcing was a huge turning point for our company. Six or seven years after that first visit to those farms in Lancaster, we were able to apply this same strategy to our supply of beef, buying straight from harvesting facilities that knew our specs, and we've never looked back.

However, nothing comes easy in this business. It took me a lifetime to get the growers we now have on our roster. At first, I was just another meat purveyor, and many growers wouldn't even listen to my requests, but that didn't stop me. As the company grew, our reputation gave more credence and power to our specs because, by then, the growers knew we were good for it, we would pay, and that's when we found our voice. Everything has a cost to it, and rightfully so. Nowadays, our growers know that if we're behind something, we will come through with our word and our money. As a purveyor, one of my jobs is to know what my customers want, get the pulse of trends to forecast what they might demand in the future, and help relay this valuable information to the growers to make it happen. Regardless of whether what we're procuring goes on to sell well or not, they are safe with us. Now there's a mutual respect and understanding. We buy from a variety of farms, both small and large, to meet our varying needs, and the lines of communication are always open between us.

With time, we have managed to develop a set of raising and finishing protocols, including the humane treatment of the animals, that must be met by the growers who work with us. Does this reduce the amount of product available to us? Sure. We need our beef to primarily grade Prime, and only around 5 percent of the beef sold in the United States falls into that category.[1] Our pool of available growers shrinks further because we predominantly choose all-natural beef, with a small quantity of organic beef and commercial beef to satisfy our customers' requests.

I began to see the word *natural* appearing on meat packaging in the early 2000s, and it caught my attention. Natural? What does this mean? Any label that's on the package or box used to pack meat has to be approved by the USDA. According to the USDA, *natural* means "a product containing no artificial ingredient or added color and is minimally processed." Right away I saw it as one of beef's first chief selling points besides its grading. As time went by, the term *all-natural* came to encompass all naturally raised beef, which basically means the cattle are raised entirely without the use of any antibiotics or hormones and no animal byproducts are added to their feed. This program, also dubbed "the never-ever policy," means exactly what it sounds like, that these particular cattle have never ever been treated with hormones or antibiotics.

Basically, if one of the animals in a grower's never-ever program gets sick, it must be pulled from that herd, removed from that paddock, given the necessary antibiotics, retagged as commercial, and added to another paddock with the rest of the commercial cattle or sold to a farm with commercial livestock. It can never ever go back to the antibiotic-free herd, plain and simple. I've witnessed this process. I've visited farms and seen growers take a sick animal out of the all-natural paddock and herd, treat it, retag it, and place it in the commercial herd. Some argue that such extreme measures aren't necessary because an animal that has received an antibiotic will have undetectable traces of it in its system after a certain number of days or weeks and should be deemed okay to return to an all-natural herd. But until there's confirmation that those undetectable traces are truly harmless, we must continue with the established parameters.

Before all-natural beef became popular, there was a big push for organic meat, which started with Whole Foods. They decided to use organic beef to distinguish themselves from other grocers, touting that it was good for us and good for the environment. General retail customers buying at Whole Foods noticed this distinction and started advocating for it outside the supermarkets. For example, a patron at a restaurant would ask the waiter if the beef being served was organic. After receiving this question several times, from different patrons, the chef would then turn to me to request organic beef. When those requests started trickling in, I took it to the harvesting facilities and asked them, "Do you have organic beef?" Posed with this query, they researched what that was and, after a while, we were able to find sources of organic beef to meet our chefs' demands. But when it got back to them, some got sticker shock. The price was almost double that of regular commercial beef. As time went by, the definition of organic changed until the USDA stepped in and began to regulate it. In order for the USDA to allow the word *organic,* like anything else that goes on a package's label, they first had to verify that the product was indeed organic. That meant sitting down and actually defining what this label would mean, waiting for the USDA's approval of that definition, and then following the necessary protocols to receive the organic certification.

Nowadays, organic beef follows the same protocols as all-natural—no growth hormones, no antibiotics, and no feed with animal by-products—but in order to be certified organic by the USDA, growers must make sure their animals are raised according to the USDA organic regulations throughout their lives and on certified organic land, which fulfills all organic crop production standards and gives the animals access to certified organic pastures.[2] For growers to go this route, they must first transition their land by not applying prohibited substances for three consecutive years. Only then can they get their official organic certification from the USDA.[3] That's part of the reason why this type of beef is more expensive: it requires quite an investment. Some growers find these initial expenses prohibitive, while others have cashed in on the fact that so few want or are able to take the big leap into organic farming.

During the big organic beef surge, we saw an equally large jump in orders and sold a decent amount of that product, until all-natural hit the market. It cost much less, and was still antibiotic- and hormone-free, so chefs began gravitating toward it instead. Since then, I've seen organic beef orders decrease and all-natural increase to the point where only about 5 percent of our orders are organic now. It all really comes down to the customers' demands. If there's ever a resurgence of organic orders in our industry, the news will trickle down to the growers, and they may find it to be just the incentive they need to transition into the organic market.

Farmers will grow anything we want them to as long as they know that once their product is ready for harvest, there will be a demand for it. It's their livelihood. If they spend their valuable time and savings to be certified organic and then come to realize there is more supply than demand for this product, they could easily go out of business. For many of the growers who want to step their enterprise up a notch from regular commercial beef, all-natural seems like the logical way to go.

Not all growers raise all-natural or organic cattle, because they must follow stricter protocols to meet the USDA's guidelines, so we work with a mix of large and small farms on a national level to procure this type of beef. If we relied solely on local farms, as some advise, we wouldn't be able to meet our current demand of feeding approximately thirty million people annually. This doesn't cancel the local growers out, by any means. We use many local farms from all around upstate New York for our supply of rabbits, goats, and lambs. And we turn to small and large growers across the nation for our supply of beef.

Taking all of this into consideration, plus our need for Black Angus beef—which does a far better job than Hereford at consistently rating Prime—really narrows down the pool of growers we can include in our program. At first, we handpicked the farms we wanted to work with ourselves; however, when our company began to evolve and grow, we realized it would be impossible for us to personally visit every prospective farm and see every animal, so we hired a beef procurer. This person is on the ground, helping us choose which farms we allow into our program, visiting them to check the animals' welfare, and making sure those growers meet our protocols and specs, comply with our requests, and live up to our standards to continuously keep our quality consistent.

Growers are remunerated based on a performance rating, which means that they get paid once the meat is harvested and graded. We'll dive into grading and labeling in chapter 4, but a key takeaway to keep in mind here is that if the animal is mistreated, bruised, or not fed properly, it will not get graded. There's no way around it. So, if a farmer dedicates two to three years to raise Prime steers and heifers and they don't get graded because of mishandling, they would be shooting themselves in the foot financially. All the money they invested in what they thought would rate Prime would go down the drain ungraded and the meat would have to be sold as cheaper commercial beef. This is actually a good thing because it keeps everyone accountable. If the growers take shortcuts on the feed, if they mistreat their animals, if they keep them in such close quarters that they're stacked on top of one another kicking left and right, at the time of harvesting, those animals will not make the cut. What does this ultimately mean? That animal welfare is not only in the best interest of the cattle but also of those who raise them.

Growers are constantly evolving with the times. Now patrons at restaurants have started taking an interest in where the beef is raised, so the new question they pose to the waiter is, where did it come from, what farm? As with the organic beef surge, the waiter asks the chef, who turns to us for the answer. When we tell the growers that the chefs want to know their names, they look at us dumbfounded but usually consent. We relay the information back to the chefs, who later come back to us with, "I can't find them. They don't have a website."

That's true. Growers usually don't have websites, they don't speak to the media, because they spend every waking hour focused on keeping their business up and running. If a chef really wants to use their farm's name on a menu, it's usually fine by the growers, but here's the conundrum for the chef: the chances of procuring beef from the same farm all fifty-two weeks of the year are slim to none. Given cattle's life cycle, farms only have steers and heifers at the right age and weight ready to sell a few times a year on average. So a chef would have to seasonally update their menu when the farms change. Some agree to this, and although this adds complexity to our process, we aim to please, so we alert them to the name change each time it comes up. However, more and more chefs have asked if they could simply use our name instead. They know we're sourcing their beef, that we are respected in the industry, and that we have certain standards and protocols we live by, so they feel comfortable using our name, which also saves them the time and money of having to update the menus every season. It's something that has started happening in the last few years, and is likely the only instance where a meat purveyor rather than a farm is listed so frequently on menus.

Meanwhile, this allows the growers to focus on raising their animals and maintaining a profitable enterprise. Do the farmers want more business? Sure, like anyone else, they want to grow their business and eventually leave it to the next generation as a source of income and as their legacy. If only it were that simple. According to the USDA's National Agricultural Statistics Service, in 2018 the United States had 2,029,200 farms, approximately

140,000 fewer than in 2009.[4] And in 2017, of those 2-million-plus farms, 729,046 were beef cattle operations.[5] That means that only around 36 percent of US farms are dedicated to beef cattle, which are intended to feed us. Why am I throwing all these numbers at you? Because I think it's important for us to realize that these families, owners of the most important agricultural industry in the United States, need our support.

The average farm in 2018 had 443 acres,[6] and the average farm's real estate value (which includes the value of land and buildings) in 2019 was $3,160 per acre.[7] This does not include other assets such as livestock. Most farmers operate on a land-rich, cash-poor business model, meaning that they invest their hard-earned money in illiquid assets (those that cannot be easily converted into cash) to help run their enterprises. Some are barely making ends meet because of the nature of the business. And don't even get me started on what happens when older growers pass away and their children inherit what is already rightfully theirs.

Yes, the estate tax, that 40 percent tax imposed on the children's inheritance. Given exemptions and deductions and work-arounds, not all farmers have to deal with this burden, but those who do are oftentimes left scrambling, having to sell off valuable equipment or a piece of land to pay the federal government, and in many cases the local government too, as many states have their own inheritance tax law in effect. The fact that it affects any farmer is egregious. Don't forget, we're talking about a tax on property and estate that have already been taxed. So, let's say that after all the deductions and exemptions, a family is left with a marketable estate value of $1 million. That means they will need to fork up $400,000 to pay the estate tax. If they have to sell land, or equipment, or livestock to cover this cost, their business will take a hit. Some don't manage to survive this hit. And that's something we as a nation can't afford, because these are the people who provide the food we eat. And while the exemption on the estate tax has more than doubled in the tax code overhaul of 2018, that is a temporary fix; it expires at the end of 2025. That leaves farmers in the same predicament because their future plans will likely go beyond this five-year mark, and who knows what will happen then.

Why should they carry this weight? Why should they have to spend thousands of dollars of their hard-earned cash on estate planning to help their children navigate the estate tax in the future? Some go as far as restructuring their business to help their children, sacrificing prosperity in the process. Why not repeal it once and for all and let them use that money to reinvest in their business and grow? If there are twenty farmers paying the estate tax and suffering the consequences in one year, then that's twenty too many. This must be fixed.

The growers are not the bad guys in this story. They work tirelessly, deal with enormous expenses, and at the end of the day, they need to make sure they can sell their product in order to survive. We should support our domestic growers rather than make sweeping conclusions suggesting that all nonorganic enterprises are factory farms and all growers are driving around in Bentleys. I know many of our nation's growers; they follow the humane and sanitary guidelines mandated by our government and spend most of their waking hours, come rain, shine, sleet, or snow, caring for their cattle despite all the presumptions they have to deal with from people who don't have a clue about how this industry works. I'm talking about real, hardworking families who have very real bills and expenses, children and relatives to care for, and who also provide jobs in their communities and ultimately feed our country. They are willing to evolve with the times because they want to help, they want to grow with us, they want to adapt to our needs. They are truly listening. The growers aren't the enemy. They are our friends, our allies, and the backbone of this industry. They should be supported and celebrated.

# 2

# THE FEED EFFECT

hat do growers feed their cattle? How do they make that crucial decision and how does that affect what eventually lands on our plates? It all comes down to choices, preferences, and resources, both on a macro and micro scale. The freedom to choose is a cornerstone of the American way of life—we consider it one of our most essential rights—and it is reflected in how our food is produced. The land, resources, and climate the growers have to contend with, together with their goals and the market's demands, are all crucial in defining their raising—all-natural, organic, and/or commercial—and finishing protocols. At some point while raising their cattle, growers have to answer the following question: Where is my beef going? Whether they aim to produce high-quality Prime or affordable commercial beef, the feed they choose from the point the calves are weaned from the cows will play a vital role in their outcome and success.

Before we get into the world of feed, let's try to better understand the animal itself. Cattle have four-chamber stomachs, the first and largest of which is called the *rumen*. That's where the main action takes place—and that's also why cattle are called *ruminants*. The rumen in each animal can store twenty-five gallons or more of material, depending on its size. It is also home to the microbes that help them ferment and digest the feed,[1] including the fiber from plants that the human stomach would not be able to process. It is such a complex system that there are ruminant nutritionists who work with growers to help each one develop feeding practices and rations that will serve the nutritional needs of their specific herd.[2] It's important to realize that raising these animals to meet our consumer standards takes a lot more than just putting them out on a field to graze. It's an intricate system with many moving parts. Most consumers are familiar with the term *grass-fed*, and many may also be familiar with *grain-finished*. Those labels refer to the cattle's food, but we must dig deeper to understand what they actually mean.

Grass-fed beef—which refers to cattle that are fed grass throughout their lives—shares the spotlight with organic beef, aka cattle raised with no growth hormones, no antibiotics, no feed with animal by-products, on certified organic pastures. Although grass-fed beef can technically not be organic if the cattle aren't raised on organic soil, the grass-fed beef we carry at LaFrieda is also organic. That's usually what our specific demographic of consumers wants, not one or the other. However, grass-fed doesn't necessarily mean that the animals have spent their entire lives grazing in open paddocks filled with lush green pastures.

First off, the USDA Food Safety and Inspection Service definition of grass-fed or 100% Grass Fed includes animals that, after being weaned from their mothers, were fed only grass. They cannot consume grain or grain by-products but are allowed to eat forage, which includes not just annual and perennial grass (i.e., those bright green blades conjured in our minds when we hear the word *grass*) but also forbs (which are basically herbs or weeds other than grass, such as brassica or legumes); browse (tender shoots, twigs, tree leaves, and shrubs); cereal grain crops in their pre-grain state; as well as hay, silage, and other roughage sources. (Incidentally, these animals will most likely also receive routine vitamin and mineral supplementation.[3]) Why all these options? Because the reality of the matter is that not all grass-fed animals can actually graze outdoors. They may be able to do so in places where large swaths of land and temperate weather allow them to roam freely outside year-round, but that is not the case in countries with less available land and varying seasons. Animals in those places have to be sheltered from extreme weather or brought in to let smaller fields rest and allow for new grass to grow. In other words, grass-fed doesn't automatically mean pasture-raised.

Something else to consider: If your product's label says *grass-finished*, this does not mean the animal was grass-fed throughout its lifetime. Grass-finished means that those cattle ate grass toward the end of their life cycle, but they may very well have been fed grain earlier.

When grass-fed beef started to become popular in the early 2000s, the go-to source was Paraguay, which is where we turned until my USDA inspector pointed out that nowhere on the boxes we received from them was a label stating that it was actually grass-fed. He asked me for documentation, so I had to get a certificate that specifically said: "Feed: grass." However, after a little more research, we found out that in the dry seasons, those growers supplemented their cattle with . . . grain! Suffice to say, we stopped buying product from Paraguay and set out to find a domestic producer to satisfy our needs instead. That's around the time I met Bill Kurtis.

Bill founded the Tallgrass Beef Company in 2005, producing grass-fed and grass-finished beef in the United States. It all began on his ten-thousand-acre farm. He was drawn to the idea of raising cattle in a sustainable way that was good not only for the environment but also for the animals and the consumers. Bill's conviction about grass-fed beef was so compelling that, when we met, he sold me on the idea practically on the spot. Having a celebrity such as Bill, who is a journalist, narrator, and former news anchor, championing this product was a fantastic idea and an excellent marketing strategy. It seemed like a win-win, so we hired a salesperson to go on the road in Manhattan with samples of grass-fed beef from Bill's farm. But the feedback we received fell far short of our expectations. The chefs simply didn't like it. I was shocked. I thought it was a surefire sale given its favorable marketing points, but in the restaurant world, it all comes down to appealing to the chef's palate, and grass-fed beef wasn't cutting it. No one really liked the flavor, and they didn't buy into the sustainability selling points, which in my business translated into low sales. Grass-fed beef is more expensive because it costs more to produce and it's sourced in smaller quantities, so when I realized my chefs wouldn't buy this type of meat, it just wasn't financially viable for me to continue carrying high quantities of it. I eventually had to call it quits with Bill. We had failed, or so I thought. What I came to realize later is that we were actually just ahead of our time.

As the years went by, the grass-fed beef sector really grew on the retail level, largely spurred by Whole Foods. I didn't realize then the power of advertising, how much money there was behind Whole Foods, and how this combination would eventually affect this part of the market. It wasn't until Whole Foods started a campaign based on their mission to offer "the finest natural and organic foods" that grass-fed began to hit the mainstream in full force. Eventually other retailers caught on, and the demand increased substantially, except in the restaurant industry. Grass-fed beef continues to be a hard sell for chefs because their number one priority is flavor. If their beef doesn't ooze sweet and juicy gusto, they lose patrons. And where does that flavor come from? Intramuscular fat.

Cattle that are 100% grass-fed certified produce lean beef. This beef lacks the delectable intramuscular fat we rave about when acquiring a high-quality cut of meat. What did you expect from a life of eating only grass? Enter grain-finished beef. Our biggest program at LaFrieda, all-natural beef, uses cattle that are mostly grass-fed and then finished with grain. Yes, cattle in the all-natural program also graze on grass for most of their lives, but in the last few months of their lives are fed grain. Why? To produce that wonderful intramuscular fat that marbles the steak on your plate and infuses each bite with just the right amount of tender sweetness. Although higher grades of beef are pricier, organic grass-fed beef also tends to be expensive due to the cost of setting up and maintaining organic soil, but that doesn't necessarily translate into higher-quality beef.

**AS I'VE MENTIONED,** animals in the industry's all-natural program are never treated with hormones or antibiotics and never have animal by-products in their feed. This doesn't mean that all grain-fed or grain-finished beef is created equal. If a herd is destined for that upper echelon of highly graded beef, they will likely be given more expensive feed. The better quality the food they eat, the better quality the food on our plate will be. Beef that is intended to be sold as cheaper commercial product doesn't follow the same protocols. Those animals usually *do* receive antibiotics in their food, because they are given lower-quality feed, which is less expensive and keeps costs down but may mess with their digestive system. That's likely where the anti-corn movement—activists against feeding corn to cattle—took root. When cattle are fed too much corn they become bloated because it changes the chemistry in their digestive system. It's like eating an entire half gallon container of ice cream. You're not going to feel all that great after you swallow that last spoonful; however, that doesn't necessarily mean you can't eat ice cream. It all comes down to the quantity and quality of what you're eating.

Corn is actually a good source of fiber and starch when harvested as silage (which, by the way, is what many grass-fed cattle eat), and the rumen microbes use the ingested grain form as a source of energy, which helps their fermentation process.[4] The key is to allow the microbes in the cattle's stomachs to adapt to the transition between grass and grain feed. If their feed is switched too quickly, it could disrupt the rumen microbes' pH levels, making them produce more acid. By the way, this can also occur with pasture-grazing cattle.[5] However, if the transition is done slowly, with the right feed mix, under the necessary supervision, the animal is able to adapt to this change and use this high-energy food more efficiently. If you've ever gone on a diet that requires you to give up caffeine, for instance, there's a transition period where you will likely not feel too great. You may have headaches and feel sluggish until you adapt to your new way of eating. It's not all that different for cattle.

I've been on a farm in Ohio with a corn grower and have seen how they turn corn into cattle feed. They don't strip the corn kernels off the cob and feed it straight to the animal as is; they dry roll, crack, or steam flake it first. These are different methods of processing the corn to expand its surface area, allowing the rumen microbes easier access to its starch, which makes it more digestible. I've held this flaked corn in my hands, and it looks just like the cereal in the box of cornflakes sitting on your counter. Standing next to one of these processing machines, I closely observed these yellow flakes rain out of the rollers warm. The farmer put a few in my hand, and I tossed them into my mouth. To my pleasant surprise, they didn't just look like our morning cereal; they tasted like it too. Furthermore, though some growers do choose to finish their animals solely with corn feed, this flaked corn is often combined with other grains to create a balanced feed mix for the animals.

So, once the cattle are transitioned from grass to grain feed, where does this grain-finishing happen? On feedlots. There, I said it. The word that makes everyone squirm. Something I hope to demystify for you now. According to the USDA, an animal feeding operation (AFO) is a facility that "confines, stables, or feeds animals for forty-five days or more in a twelve-month period." A concentrated animal feeding operation (CAFO) needs to meet the criteria for an AFO and confine more than one thousand beef steers.[6] These AFOs are what we identify as feedlots or feed yards, the last stage where the animals are grain-finished before they are sent to the harvesting facility.

In the 1950s the demand for quality beef began to skyrocket, pushing growers to innovate and find ways to supply their customers with what they wanted. By the 1960s, they managed to increase their beef production considerably due to the implementation of feed yards, which began to produce heavier animals at a faster rate because the grain provides the fat that turns into the sought-after marbling in beef graded Select or higher. And that's how the feed industry was born. By implementing feed programs, the resulting quality of the beef became more consistent.

The industry's efficiency only evolved from there, adapting to consumers' needs as they grew more attentive to beef quality in the 1980s and '90s.[7] That's also the approximate time animal activist organizations, such as PETA, started circulating videos from undercover investigations in the beef industry, depicting feed yards where animals were bunched together in overcrowded pens, housed in unsanitary conditions, stepping on and kicking one another just to get to their food. Those widespread images created a huge stigma around feedlots in the public eye that endures to this day, even though those archaic methods have vastly improved in part due to all that negative publicity. Their progress also had to do with the growers themselves and the profitability of their business.

As we explored in the previous chapter, if an animal is kicked and bruised, it's to the detriment of the grower because he or she will get paid less if the beef doesn't make the grade. So it doesn't behoove the growers or feed yard owners to mistreat their animals. Usually, they have employees checking up on the herds on a daily basis, tending to their needs, with vets and rumen nutritionists at the ready to make sure their stress is minimal and they are healthy. Why all that fuss? Because that will ultimately produce better-quality beef. Most people don't realize this connection and what a big incentive this is not only for the growers but also for the feed yard industry to treat their animals humanely. Behind the scenes they are constantly upgrading their practices and striving to make them as efficient and humane as possible. This keeps their customers happy and their families in business.

The beef industry evolves alongside society. Consumers nowadays not only want to feel comfortable with what they're eating but also take an interest in the environment, which leads us into tricky terrain. Let's start with what we do know, the thing that has everyone up in arms about cattle: methane production. When cattle eat, they ingest their feed, then regurgitate it to chew the cud, and in this rumination process, they burp the gas methane. Incidentally, methane doesn't only come from cattle—it also comes from the production and transport of coal, natural gas, and oil, among other sources.[8]

Now, when it comes to organic matter related to cattle, here's another fact many people don't know: it's mainly cattle burps and not farts that emit methane, as does their manure, and it's been going on for as long as these animals have been roaming the earth because their digestive system hasn't changed. Furthermore, cattle ingest the same amount of carbon through their food (because plants use carbon dioxide during their photosynthesis process) that they release when they burp.[9] This is called the biogenic carbon cycle.[10] If this is the case, then, unlike fossil fuels, which actually produce new $CO_2$ and release it into the atmosphere, cattle are not creating any new $CO_2$.

With time and experience, I've learned to become a forensic reader when it comes to the media and any type of data that receives a ton of attention. Regardless of whether it's ultimately true or false, my initial reaction is always skepticism. Who provided the data? Who funded it? Who are the scientists making the claims? Where do they work? Who pays them? I ask so many questions it may dizzy some people, but my only intention is to get to the heart of the matter.

Case in point: the 2006 Food and Agriculture Organization of the United Nations report titled "Livestock's Long Shadow," which concluded that livestock (and here we're talking not only about beef cattle but also dairy cattle, chicken, pigs, and so forth) produced more greenhouse gases than the entire global transportation system.[11] Everyone latched onto that line and ran with it, so much so that it's still seeping into reports today, almost fifteen years later. But it wasn't accurate or a fair comparison. The FAO used a life cycle assessment to create that specific report, which means it included every minute detail regarding what it takes to produce livestock, from soil to feed to raising, finishing, and harvesting the cattle, to the refrigeration process and every other step it takes to bring a cut of beef to your plate. Yet the percentage they used for the global transportation system's greenhouse gas emissions as a point of comparison did not stem from the same type of assessment. They didn't measure transportation's environmental footprint from start to finish—they only considered its tailpipe emissions. No one sat down and calculated what it took to make and deliver each vehicle part, the emissions produced at the factories that build these vehicles, the transport of these trucks, planes, and automobiles to their final destinations, and the emissions they produce once they're on the road. No one said a word about this until a group of scientists led by Frank Mitloehner, a professor and air quality extension specialist at University of California, Davis, noted the enormous discrepancy.

It was such a glaring inaccuracy that the FAO themselves backtracked and admitted their mistake, but by then it was too late. Subsequent reports have been largely overlooked because they no longer carry the stark impact certain sectors need in order to advance their own interests.

According to the Environmental Protection Agency, in 2018, "The largest source of greenhouse gas emissions from human activities in the United States is from burning fossil fuels for electricity, heat, and transportation."[12] When they break it down, the order basically amounts to transportation (28 percent), electricity (27 percent), industry (22 percent), commercial and residential (12 percent), and agriculture (10 percent). Of the 10 percent of greenhouse gas emissions that agriculture represents, less than 2 percent come from beef cattle emissions.

If cattle were eradicated, it would put only a 2 percent dent in greenhouse gas emissions, yet it would disrupt the environment in a way that could lead to another Dust Bowl like the one that swept through the Southern Plains region in the 1930s. Why? Because cattle grazing actually benefits the environment in many respects. Having herds of ruminants grazing on our land is nothing new; it's been happening for hundreds of years. According to the U.S. Fish and Wildlife Service, in the 1500s, thirty to sixty million bison wandered around North America,[13] playing a crucial part in the ecosystem. As cattle roam the fields, they help decompose dead leaves, their hooves push seeds into the soil, and they fertilize this soil with their urine and manure, which keeps the soil healthy. There's an entire movement known as regenerative agriculture[14] that posits that certain farming and grazing practices can reverse climate change by working to restore soil biodiversity and carbon sequestration. Deep-rooted grasses, which cattle graze on, actually prevent soil erosion and sequester carbon. However, this doesn't mean grass-fed cattle are the answer either. Having animals graze is a good thing. And most cattle graze on grass for more than half their lives. However, grass-fed cattle ultimately require more land and extra months to reach their marketable weight, which means that they will be around burping methane for a longer period of time than grain-finished meat.

It takes at least twice as much time to grow grass-fed beef that is profitable to the farmer than it does to grow grain-finished beef. Growers need their steer and heifers to reach a certain weight before they can sell them for a profit. Since cattle raised solely on grass take longer to develop, the grower must raise them for a longer period of time to reach market weight. There's an inevitable environmental impact as a result of all that goes into extending their life cycle. A Harvard study conducted in 2018 found that if the United States transitioned to exclusively grass-fed beef, the country's cattle herd would need to increase 30 percent to meet the current demand, which would lead to a 43 percent increase in methane emissions from that sector. However, eventually, fewer cattle could be raised in the United States due to lack of land available to host so many heads for a long time. Fewer cattle would mean less food for our ever-growing population. Furthermore, decreasing the amount of available beef in the United States would increase its price.[15] Additionally, there are studies that concluded that the smaller land footprint and high-energy feed used in feed yard systems are associated with lower greenhouse gas emissions.[16] And none of these analyses explore flavor, but older beef is inevitably tougher and less tasty.

What's better in the end for consumers, for the environment, for the world? I honestly don't know. From what I can tell, the research is still in its infancy. We need more numbers, more proof, better comparison studies. Only time and objective studies will lead us to the answers we are all seeking, and that includes the beef industry, which is always looking to become as efficient and streamlined as possible. At the end of the day, feeding the rising world population will continue to largely fall on our shoulders.

When people traffic in unfounded presumptions about meat, it bothers me. I can't help but envision the families caring for those animals, taking all the necessary precautions, complying with USDA regulations, going above and beyond to deliver a safe food source and keep one of our last natural resources thriving, only to be lumped together with those who are flouting the rules.

Trials at UC Davis have aimed to find ways to lower methane emissions from cattle to make the industry more sustainable. Their initial findings indicate that by using 1 percent seaweed in the cattle's diet, methane emissions could be reduced up to 60 percent.[17] Meanwhile, some dairy farms are feeding their cattle under tarps to trap the methane they emit and turn it into fuel[18]—capturing these emissions and transforming them into renewable energy. There are other studies that conclude that cattle's methane output could be genetic, which means that certain herds might emit less than others, so it could come down to breeding the latter out to be left with less gassy animals.[19]

Clearly, there are many possible solutions. I find this quite promising and inspiring, especially with the predictions about how the population will increase by 2050. The beef industry has managed to produce more beef per animal as it has become more efficient over the years, which means fewer cattle produce as much or more food than a few decades earlier. Again, all of this is promising news.

I care about the earth. I have two kids—of course I worry

about the environment and I want them to have a planet that they can enjoy as much as we have. Fossil fuels, which continue to be the main offender, should be where we should focus our energy while the beef industry continues to innovate and improve through technology and management. Something as simple as wasting less food would help. Remember, twenty steers and heifers can feed several hundred people, and food security is still an issue in our country as well as around the world. Doing away with a critical food supply isn't the answer.

I'm sometimes painted in the media as being against grass-fed beef. During a live interview on Bloomberg, I was suddenly hit with this question: "Why is it that you're withholding the fastest-growing sector of beef from the American public?"

"What sector is that?" I asked, puzzled.

"Grass-fed beef."

Someone did a little research and found out that at some point I mentioned I didn't *like* grass-fed beef from a quality perspective. And, no, I don't, because it doesn't have the marbling needed for it to be the Prime cut I enjoy. But that doesn't mean I withhold it from my customers. I'm not sure how that conclusion was drawn. I always have and always will give my customers what they ask for, regardless of my own preference. However, when someone assumes that my preference is reflected in what I sell, I have no choice but to correct them. Grass-fed is also part of my business, albeit a smaller percentage, and not because I'm against it or withholding it, but because that's what our customers demand. I'm a purveyor, so I have to source the best product in each category and supply it. I listen to my customers—they are my industry compass. If they demand more of one style of beef, I will do everything in my power to get it to them.

The time it takes for general demand to make it to the growers and the time it takes them to turn around and be motivated enough to grow, say, grass-fed beef, or more Prime beef—well, it doesn't happen overnight. Taking into consideration the USDA approvals, and the actual raising and finishing of the animal, means three years as a minimum for me to get that first pound of whatever is in demand, and then I have to hope that this trend didn't end by the time it hits the market. I can't tell my grower that he or she isn't going to get paid what was promised. So how am I withholding any type of beef from the American people? Right now, restaurants tend to prefer my all-natural program, while grass-fed beef is more popular in retail. And as a consumer myself, sure, my preference is all-natural, grain-finished, Prime beef. But if there's an increasing demand for grass-fed beef, then grass-fed beef is what people will get.

And when it comes to the environment, let's not be blinded by politics or extremists. Look at the evidence and ask the questions. As in, what does one private jet do to the environment and why do green politicians fly on these private jets? Is it a coincidence that the up-and-coming plant-based companies are backed by influential tech leaders? Is that why beef so often ends up being the environmental scapegoat? I ask these questions because I really want to know. It's baffling at times, I understand, but if we don't get better, more transparent answers, we won't know how to truly help sustain the environment while also feeding the millions of people who rely on this source of food.

At the end of the day, farmers, growers, harvesting facilities, meat purveyors, restaurants, and retail companies all have the same objectives. We strive to stay profitable, to meet our consumers' demands, to take care of our families, and to survive and evolve with the times. Livestock health is as important to the growers as it is to the consumers because it translates to healthy food on our plates. We're in this together. That's why it's so important to continue to have these conversations, no matter how uncomfortable they may be, to stay informed, to keep each other accountable, and demystify the misconceptions surrounding beef.

# 3

# HARVEST TIME

Once I started cutting meat with my father and grandfather on Bleecker Street, I was absolutely hooked. I loved every aspect of it. However, the last thing my dad wanted was for me to follow in his footsteps. My grandfather had taken over the business from his father and was proud to call himself a butcher, but my dad would've been happy doing something else. He had other dreams, like working with cars. Music was another interest that moved him; he was a drummer in a band that performed at basketball game halftime shows in Madison Square Garden and traveled upstate during the summer to play in those bygone resorts everyone used to flock to—the Hamptons of yesteryear. He even landed a gig with Tito Puente once. My dad loved music so much that when I was a kid, he hoped I might follow in his musical footsteps rather than go the butcher route. I patiently sat at the drums he had lovingly set up for me at home, but it was no use. I just didn't get it. My coordination was off and my heart wasn't in it.

Who knows what road my dad would've taken had it not been for his family mandate? No matter how many dreams he had, as his father's only son, the decision was forced upon him. It was the old-school Italian way of doing things: you had to join your father when it came to the family business; there was no way around it back then. With years and experience, and I suspect through the bond he shared with his dad, my father eventually wound up liking the butcher's way of life. Yet he was dead set on deterring me from that path. Determined to provide me with the choices that had been snatched away from him, he worked me hard on those days when I went to the shop with him, making sure I did all the menial, tedious tasks that he secretly hoped would discourage me from becoming a butcher just like him and my grandfather. Yet, no matter what he did, I didn't budge: this was what I wanted to do. That's when it hit him: it's one thing to cut beef in the shop, it's quite another to see what happens before those primal cuts hit the butcher table.

A few weeks earlier, my dad had taken his brother-in-law to a harvesting facility over in New Jersey, and after my uncle saw the first animal go down, he turned pale white, exited the building, and tried to buy all of the remaining cattle on the lot. Bingo! My dad was convinced that if he brought me there too, the somber experience would set me straight and scare me away from the family business for good.

I was around twelve or thirteen and I was obviously aware that the primals we handled at the shop originated somewhere, and I could imagine the likely method. I knew these facilities existed—I had heard stories from my dad—and, as I understood it, this part of the process was necessary for our existence. After all, we've been killing and consuming animals on a regular basis for more than two million years.[1] So across the Hudson River and into New Jersey we went. After greetings were exchanged and procedures were explained, we entered the facility, and I simply stood there and watched from start to finish. No matter what role you have, a harvesting facility is not a happy place, but the color didn't drain from my face and I didn't feel squeamish. It was certainly a sobering moment, but I took it as a learning experience.

Once the facility employees had killed the animal and opened it up, I intrepidly stepped up to the carcass and carefully placed my hands inside it, feeling each organ, identifying each part of the animal I had only known up until then as primals on a butcher table. Who knows, maybe it was all the years I had spent fishing on my dad's boat during those long, hot summer days. I actually learned how to skin and debone a fish long before I started cutting beef, so I was already familiar with the insides of an animal, but I had never before seen such a large one. When my dad returned to check up on me, assuming I would be long gone by then just like my uncle, he was taken aback by the scene. I was still standing there, unafraid, my curiosity piqued. His plan had completely backfired and only deepened my interest in the business.

Harvesting my own food, be it from fishing or later on from hunting, has always been part of my life. I come at it from a different perspective than a person who has never had to break down a primal. Now, doing it in the most humane way possible is always a priority and something that became mainstream in my line of work due in great part to the tireless research and innovation of livestock industry icon Temple Grandin—someone I had the pleasure of meeting a few years ago. With her degree in animal science and her keen observations of cattle behavior, Temple set out to help harvesting facilities put into practice a more humane slaughtering system, one that would keep the animals calm, allowing them a gentler ending. In turn, this would help the facilities comply with the Humane Methods of Slaughter Act,[2] which was signed into law in 1958, expanded in 1978, and made fully enforceable in 2002 (by the way, it does not apply to chicken or other birds). As Temple famously has said about animal slaughter, "Nature is cruel but we don't have to be. We owe them some respect."

By patiently observing cattle behavior, she realized these animals don't like corners, which led her to create curved holding pens and loading chutes. Then, in the mid-1980s, Temple developed what would come to be known as the double rail restrainer conveyor. In this system, cattle are gently ushered in single file onto a curved loading chute and up an entrance ramp, both of which are nonslip to ensure they don't lose their footing along the way, which would inevitably cause their nerves to spike. Then they are eased onto a false floor that becomes a conveyor with adjustable sides that rise up for the animals to straddle, comfortably fitting their brisket and carrying them the rest of the way. Meanwhile, a solid hold-down rack above the animal runs the length of the conveyor to keep it from seeing people or catching sight of any other unnerving activity.[3] Temple's brilliance drove her to apply behavioral principles to control the cattle instead of using brute force, permitting them to ride through the system with a sense of tranquility, which in turn eased the subsequent stunning process.

After her ingenious findings went public, many harvesting facilities retrofitted their systems to comply with her circular design, and she became the godmother of humane slaughter. Furthermore, this conveyor restraining system made the end product's quality even better. If an animal is stressed out, anxious, or terrified, that fight-or-flight reaction will naturally blast adrenaline through its body, tensing it up and resulting in tougher beef. This no longer happens in facilities that use Temple's system and a stun machine.

It bears repeating: the meat industry needs to be efficient and ever-evolving in order to survive. It's the nature of the business. That's why we pay such close attention to what consumers want and need. When harvesting facility owners noticed that their customers cared about the humane treatment of animals, many of them naturally upped their game. How did this message get to them? It's the same trickle-down effect we've discussed before: the growers learned about the demand for all-natural or organic or grass-fed beef from the butchers and meat purveyors when they began to receive these specific requests or inquiries from restaurants and markets, because their customers increasingly began to ask how the animal was harvested and if its welfare was taken into consideration.

Stunning technologies were introduced in the first half of the twentieth century, enforced in the United States through the Humane Methods of Slaughter Act, and really gained traction as an industry standard at the start of this century. Stunning methods basically consist of an electrical current administered to the animal, rendering it unconscious and making its final stage pain-free. I always say, if it's my time to go, just stun me, because you're out cold and don't feel a thing. One bolt and it's done. As my friend the late food writer and historian Josh Ozersky said after touring one of the industry's humane harvesting facilities, "The process is a somber one. No one takes this lightly. It's sobering, but not a horrifying one. It's not the horrific, violent scene that I always thought it would be."[4] I couldn't agree more. Stunning is not only more humane, it produces better-quality meat because it helps avoid the adrenaline that surges through the animal's body with slaughtering methods that don't include stunning.

Those gruesome clips that have circled the world—and are embedded in our minds and recalled anytime we hear the word *slaughter*—are actually misleading. Yes, they're intense and bloody, but they don't represent most US harvesting facilities. Those undercover clips were mostly filmed at religious plants, where stunning is purposefully avoided, causing the scene to be more violent in nature. If you look closely, you will also likely see someone in the lines wearing religious attire—that's another key giveaway. For example, kosher harvesting requires a neck incision to cut off the animal's blood supply, which means it will take a few excruciating minutes to bleed out and die. It's no wonder everyone is up in arms when they see this. However, even religious harvesting facilities are keen on evolving their methods to improve animal welfare. They are exploring new technologies in the hopes of being able to apply alternatives to stunning that will ease the animal's suffering while still complying with religious laws.

**AFTER BREAKING FREE** from the Fourteenth Street Meat Market and making contact with growers, my dad and I began to travel to different harvesting facilities to see what their process was and to give them our protocols and requirements for continuing to expand our supply chain. It wasn't an easy conversation to have because these companies deal in millions of pounds of beef per week and there we were handing them a list of our specifications, including one about the eye of our veal having to be six inches or more in diameter, knowing full well that we were barely a tiny fraction of their business. The thought of someone holding up a ruler on the harvesting floor to comply with specifications for these nobodies, well, let's just say we were initially received with astounded laughter all around . . . until we told them we were willing to pay in advance to show we were legit. That's all it took to quell those chuckles and get down to business. Money talks, especially when you're trying to establish your reputation.

Harvesting facilities need to make sure they get paid, just like everybody else. If they process animals and then have nowhere to send the primal cuts, what are they supposed to do with the beef in their facility? Not moving their product is just not sustainable, so they need to make sure they work with people they can trust to take their product off their hands in a timely manner.

It took a few years to establish these relationships, but our solid reputation of on-time payments and reliability grew stronger over time. Now we pay for the harvesting facilities' beef within seven days of its leaving their building. To put into perspective what this means for us: It might take up to four days for the product to ship from the harvesting facility and reach our warehouse, and three days after that, the payment is cleared and in that business's account. That seven-day window is challenging for us because the beef, which arrives a few days after leaving the plant, then requires another few days to get cut, packed, labeled, shipped, and delivered to our restaurant clients. Then it takes restaurants an additional forty to forty-five days to pay us. This means that I basically get paid around a month and a half after my initial payment to the harvesting facility went out. It's not easy, but it's been our steadfast policy since the start, and that's part of what keeps our sources and clients happy and gives us the power to request the specs for our programs.

We went through several harvesting facilities before we landed on the ones that we felt were the best fit. It wasn't just about proving that we could pay on time; we also had to show them that we'd be steady users, clients who would order product more than once a month. Thankfully, I was already elbow deep in growing our family business and procuring new customers, which eventually translated into heftier beef orders, but it required patience and time, years really. You can't go into a harvesting facility and tell them you can only purchase a fraction of the carcass, like the rib eyes and the strips. What are they supposed to do with the rest of the animal? When our customized burgers became popular among our clients and our ground beef program began to take off, that's when I was finally able to go back to my harvesters and order the whole front of the animal, all the way to the sirloin and even some of the hind leg, and that's when they knew we were here to stay.

Funnily enough, growing interest in where food comes from even put harvesting facilities in the limelight. It all began with a great marketing campaign initiated by restaurateur Steve Hanson and chef David Burke, who in 2006 bought a $250,000 Black Angus bull with the idea of breeding their own flavorful, marbled herd of hundreds of cattle at Creekstone Farms. It was a great story that got a ton of media attention, but there was no way a restaurant group would be able to use the amount of beef they set out to produce. And where did they plan to offload what they couldn't put to good use?

In the meantime, we heard through the industry grapevine that Hanson and Burke weren't on great terms with Creekstone, so when they finally called it quits, we stepped in. We called Creekstone and spoke to the owners about taking over that specific program, and things only grew from there. My cousin Mark Pastore ran with the marketing idea and turned Creekstone Farms into a household name. Restaurants loved it because they could easily add Creekstone's name to their menu without having to worry about changing it seasonally since they always provided a steady stream of product. How could that be? All growers know that cattle are sold for harvest at certain times of the year, following their life cycles. But Creekstone Farms is different because . . . it isn't a farm. I've explained this time and again, but restaurants and customers alike seem to turn a blind eye to the fact that this is actually a harvesting company; in other words, a kill facility. That's right, restaurants are basically listing on their menus the slaughterhouse where their steak was killed (not raised), but since the name has the word *Farms* in it, people are none the wiser, or they don't want to be. It's genius. My Mexican business partners' marketing team once called me thrilled with the news that they were going to fly up and get a tour of Creekstone Farms, and I couldn't help but quietly chuckle in disbelief.

"They're not going to let you tour the kill facility," I said.

"No, we're going to the farms!" they replied with excitement.

"What farms? They don't own any farms."

They had no idea they were about to go visit a harvesting facility. What a wake-up call that would've been—although they rarely let anyone tour the kill line. I don't own any farms, but I obviously know farmers, so I sent them to see one of these growers instead.

Our relationship with Creekstone has changed somewhat in the last few years because their focus seemed to shift to the dollars and cents of their business, and we thrive with partners who share our passion for the industry. Then they were bought by a Japanese company. After a few disappointing run-ins, we began to reduce our usage of their product and rev up orders from our other sources.

All of the harvesting facilities we work with are selected after we've visited them and ascertained that their business fits with our programs and protocols. That's when we feel comfortable enough to strike a deal. Honestly, to this day, that conversation about our specs is still a little hard for me to have because I'm highly aware that, even though I purchase much more beef than when I started, I still take only a fraction of the harvesters' output. However, earning this industry sector's trust has given way to a great reciprocal relationship that allows us to continue to conduct our business as seamlessly as possible so as to get whatever our clients need while also always coming through for our harvesters and growers. In the end, we're all in this together, aiming for common goals: to keep feeding people while staying current with trends and requirements and running things smoothly and efficiently.

Of course, as with some farmers and processing plants, there's still plenty of room to grow and improve. Growers who eventually want to sell their product to a meat purveyor, like myself, who supplies restaurants, need to transport their cattle to USDA-regulated harvesting facilities to comply with federal mandates. However, a seemingly simple process can sometimes become quite an odyssey for farms that aren't near such a location. When growers don't have access to USDA-sanctioned harvesting facilities, they're not completely screwed. They still have the possibility of using local facilities to sell direct to butcher shops, but that means that they lose out on selling meat commercially across state lines, so their reach is greatly reduced. Many growers find that they have to drive for hours on end, sometimes to other states, just to get to one of these USDA-regulated facilities. This in turn increases their processing time and costs. For growers who market their beef as local, that also means that they not only have to transport their cattle to these faraway facilities but they also have to drive the product back to their area to sell it locally. As long and tedious and impractical as this may sound, many prefer to do it to ensure that their animals are harvested in a facility that they trust will follow the necessary humane protocols. These growers didn't spend their animals' entire life cycle caring for them only to have them mistreated at the very end.

The Food Safety and Inspection Service is aware of this issue and, in order to help these farms, they have created mobile slaughter units to bring USDA-approved harvesting facilities closer to the growers in need. However, it's not the easiest thing to carry out because each unit must still comply with the same USDA regulations as those in permanent facilities, which means there must also be one USDA inspector assigned to each unit to verify the harvesting process. Currently there are only nine of these mobile units in all of the United States—hardly enough to go around—and they can process up to ten animals per day.[5] Just to put this into perspective, larger harvesting facilities can process more than one hundred heads of cattle per hour. Additionally, these units must have enough space and height to kill the animal, lift it by its hind leg, hang it (per safety regulations, the beef cannot touch the floor), break it into primal cuts, and carry this all out at low temperatures. I find it hard to imagine that this can all be done while adhering to humane protocols, but I understand the need the small growers have for these units. Without them, processing their product can become prohibitively expensive, hindering any real chance of growth and expansion.

A few years ago, the New York state government asked if I would consider taking over a harvesting facility upstate so that it could remain available to the small farms in that area. I was mulling it over, and even traveled to check it out, crunching the numbers and researching the logistics, but before I could respond, the government had already recommissioned the property to someone else, for a completely different purpose, dealing a low blow to the local farmers revered by so many residents.

In the long run, the USDA harvesting facilities are what keep America's most important agricultural industry safe. We don't want to go back to the old days of trichinosis in pork and *E. coli* in beef. Meat is one of the safest things you can buy now because of federal regulations. Just think of the hundreds of millions of meals eaten daily without devasting foodborne illness.

**UPHILL BATTLES ABOUND IN THIS INDUSTRY,** which is probably one of the reasons why my dad adamantly discouraged me from joining it, but I take great pride in how we've confronted these challenges. Did you know that practically all of the beef animal is used when it is processed? Around 60 percent is harvested for food, and the other 40 percent becomes by-products that most of us use on a regular basis. For example, the animal's pancreas is used to make insulin injections for diabetics; the adrenal glands can be used to make steroids; the cartilage helps make osteoarthritis medicine; the lungs are used in blood thinners; the fat is turned into tallow and used in some toothpastes, soaps, creams, and cosmetics, as well as in hydraulic brake fluids and antifreeze, and is even being tested to possibly be used as a biofuel.[6] From the hide comes the collagen used in plastic surgery, as well as the leather used to make basketballs, baseballs, and footballs. Other by-products include household and office items like dish soap, glue, pet food, and gelatin, as well as clothing and furniture.[7] That's the definition of efficiency right there.

# 4

# THE IMPORTANCE OF GRADING, LABELING, AND TRACEABILITY

From the late nineteenth to the late twentieth century, beef arrived in New York City through the Fourteenth Street Meat Market. The meat from every farm in any part of the state or country that shipped to the city landed in that West Side enclave. When I started working with my dad, many company owners still illegally stamped superior grades on their beef to turn it over at a higher price point. This was no joke; the consequences were severe. The last person I know who was caught had to do eight years in prison. Legal repercussions aside, this meant those stamps signified absolutely nothing to us at the time; they couldn't be trusted. It all came down to being well versed in beef and its grading system and learning how to identify the best-quality cuts amid the market's grinding fast-paced mayhem. It's not something that can be learned in a classroom; it really comes down to on-the-job training. And this is true for meat inspectors and graders too. No matter how much training they get, only time on the floor and on the line, evaluating and grading actual carcass after carcass, will help them develop the keen eye they need to do that job.

Now more than ever, the USDA inspector and grader's role is key because people are paying closer attention to and want to feel comfortable with what they put in their bodies—it's not just about animal welfare, it's about safety too. Take water, for instance: some people won't be caught dead drinking water from a fountain, but they'll happily stride into a store and buy their favorite brand of bottled water because they see it as a trusted source that will deliver a healthy, regulated, tested product. The same goes for beef. In addition to being able to rest easy with the humane treatment of the animal on our plate, we all want to feel like what we're eating is safe. That's why beef that is harvested for commercial use must follow strict USDA regulations. But humane treatment and safety aren't our only priorities. It's flavor that makes you return to your favorite restaurant, buy specific cuts from your market or butcher, or go back for seconds at home. Flavor, flavor, flavor. In the beef world, the USDA grading system actually grants us the opportunity to clearly identify the tastiest meat before it even reaches our kitchens, let alone our taste buds.

USDA grades originated in the early twentieth century to essentially help a struggling agricultural sector better market its meat and help streamline its quality. The grade standards were published and promoted by the Secretary of Agriculture in 1926 as the Official United States Standards for Grades of Carcass Beef, and in 1927 the USDA's voluntary beef grading and stamping service came into existence.[1] Since then, it has remained voluntary, but the descriptions of these standards have been amended several times to improve beef consistency and keep up with customers' preferences and the industry's evolution. Aside from determining the beef's flavor and tenderness and providing this valuable information to customers, grading also helps growers ensure that the quality of their product directly correlates with its pricing. When their beef receives a higher grade, their product goes up in value, which incentivizes them to continue applying the raising and finishing protocols that prioritize animal welfare and render succulent beef.

It all begins at the harvesting facility, where USDA inspectors and graders evaluate beef carcasses and assign corresponding grades based on their final assessment. Actually, the USDA inspector first steps in prior to harvesting to perform antemortem inspections of the cattle in the facility's holding pens. They need to certify that the animal is in good health before it is killed. Animals that are sick are removed from these pens and euthanized. The same goes for the downers, the ones that have been injured and are unable to stand on their own. The rule of thumb is that the animals must be able to walk on their own into the processing chute and entryway. If they can't, if they fall down and can't get back up, or if they die in the process, then they have to be removed from the pen immediately and are either incinerated, buried, or rendered, depending on the injury or illness that caused their demise. It's a terrible waste, but processing sick animals is out of the question.

Once the animal is harvested, its carcass is split down the back and examined by the USDA inspector to verify that there is no trace of any type of illness or contaminants that could threaten consumer safety. This is not the only line of defense. The carcass goes through other protocols that target harmful bacteria, such as hot water and organic acid rinses. Additionally, all specified risk materials (SRMs), which the USDA considers "inedible and prohibited for use as human food,"[2] are removed from each carcass—this includes the distal ileum from the small intestine and the tonsils of all cattle regardless of age. And from those thirty months or older, it means the skull, brain, trigeminal ganglion, eyes, dorsal root ganglia, and spinal cord,[3] a requirement that was spurred by cases of bovine spongiform encephalopathy (BSE), which cropped up in North America in the 1990s and 2000s and caused mayhem in our industry. I'll get to the root of this problem later on, but suffice to say that I don't buy beef past thirty months of age not just because of the quality, but to avoid the risk of coming across this disease or any others associated with older cattle.

Once the federal health inspection is finalized, the carcasses are sent to a cold room to be chilled for twenty-four to forty-eight hours before they are rolled toward the USDA graders. As each one reaches the grader, it is split between the twelfth and thirteenth rib bone (in other words, the rib and loin meat), in a procedure called *ribbing*, which must be executed flawlessly so that the carcass is eligible for grading. The cut is done in that spot because consensus in the industry points to that rib eye area as being the most consistent part of the animal when it comes to color, texture, and firmness, to perform this grading evaluation.

The graders initially must determine the animal's sex, age, and hide color, as they all play a role in what grade that beef eventually receives. Steer and heifers under thirty months of age are not only healthier but also render more palatable beef than older cattle do, and cow beef is more flavorful than that of bullocks or bulls. Good beef is under thirty months, but the sweet spot to me is twenty-two to twenty-four months. To assess their age, graders evaluate the amount of cartilage found in the vertebral column on top of the split chine bones. In other words, they take a look at the cartilage to see how much has turned to bone. If there is more bone than cartilage, they know they're dealing with an older animal.

In 2016, documentation of actual age and dentition (that is, the analysis of the arrangement, development, and number of teeth to pinpoint the animal's age) was added as a tool to help determine the animal's maturity.[4] The hide color is also key—if it is found to be 51 percent black or higher, the carcass is stamped with a letter *A* to signal it can be labeled Black Angus beef, which usually grades higher than other breeds.

Once all of this is done and ready, the graders assess the animal's fat distribution (aka the degree of marbling), meat color, texture, and firmness, which can all be determined by evaluating the rib eye muscle area that was exposed when they ribbed the carcass. One of the best ways to determine a carcass's grade is to look at its intramuscular fat distribution. These are the flecks of beautiful white fat that run through the inside of the muscle. The higher the degree of marbling, the higher the grade, and the higher the payment for that beef. White fat denotes better-quality beef than yellow fat, though when it comes to the USDA, they focus on the color of the meat rather than the fat. A light red bodes much better than a dark red. The grader will also be on the lookout for fine texture and degree of firmness. Each one of these characteristics results in quality grades that ultimately denote the beef's juiciness, tenderness, and flavor. Graders also assess the animal's yield, as in the amount of usable red meat versus fat and muscle content. Black Angus's yield is great because the outer layer of fat isn't as thick as, say, in a Holstein. And the Angus's rib eye is much larger.

It used to be that graders would take their position on the line with a small stack of photo cards in their hands illustrating the different beef grades. After splitting the carcass and ribbing it, they'd hold the card up that they felt best matched their grade assessment to see if it was a close counterpart, and that would help them reach their final decision. Thankfully times have changed. Now the grader uses a digital instrument to measure the rib eye area, which was approved for use in August 2001 and in 2009 was upgraded to also help assess the degree of marbling. It captures a digital image of the rib eye area and uses computer software to measure the yield and quality grading factors (area, marbling, color, texture, and firmness) to help determine what grade that particular carcass will receive. However, graders also inspect it themselves and input the corresponding grade into the system by hand, and they have the power to overrule the instrument's decision if they think it is not functioning properly. It's such an improvement from the photo cards, where so much money hung in the balance depending on the grader's subjective view and opinion.

The growers and harvesting facilities get paid after the meat is graded, which discourages them from taking shortcuts and incentivizes them to treat their animals humanely. Since so much is at stake, a plant employee usually stands on the line next to the grader, and sometimes a "Hey, that's Prime!" "No, that's Choice!" argument may ensue. Graders are supposed to be unbiased evaluators, calling it as they see it because their salary does not depend on the grades they dole out. When in doubt, the grader will push the beef to a third rail and get back to it later to avoid holding up the line. The USDA grader is also the one who evaluates whether the carcass meets certifiable verifications for specific programs, such as no use of hormones, raising and finishing claims, organic, and so forth. If they verify that these claims are accurate, they add an "accepted as specified" certification stamp to the carcass.

It may seem like a long process, but it's actually quite a fast-paced environment, where everyone on the line fills a specific role and gets the job done quickly and efficiently so as to get through as many animals as possible in a day. Once the evaluation is complete and the grader has made his or her final decision, the carcass is officially rolled with a quality grade stamp made of purple food-grade ink, which is really just concentrated grape juice, stating the beef's grade (Prime, Choice, Select) and yield number (with 1 designating the highest red meat yield and 5 the lowest). This number helps the growers and buyers at purchase time, but makes no real difference to the palatability of the beef from a consumer's perspective. This is why consumers need only focus on the quality grade shields adhered to the beef's packaging as symbols of a safe and premium product.

Correct grading is crucial not only to the consumer but also to the survival of businesses like my own. Last year a company in Mexico received a shipment of my product that was not graded correctly and mayhem and miscommunication quickly ensued. The buyers immediately thought, *Oh, of course, Pat LaFrieda is saving all the good stuff for America,* which was ironic because that's exactly the industry trend I wanted to help *end*. For years, Mexico had been robbed of good product, and I had set out to change that. I always like meat to come through my facilities so that I can verify firsthand that everything is in order, but in this particular case, this company had requested it to be shipped straight from the harvesting facility so they could save a few cents. The USDA grader who normally worked at that facility wasn't there on that particular day, so it likely came down to that push and pull I mentioned earlier between a less experienced grader and the savvier plant guy who got away with this particular beef being graded as Prime when it was not. My intention was never to send them below-average meat, so when this was brought to my attention, I was furious. I told them that if they received anything that didn't meet their criteria, they should send it back. Finally, the issue was remedied and the relationship was saved. But it was a close call.

**BEEF CAN FALL INTO** eight different quality grades: Prime, Choice, Select, Standard, Commercial, Utility, Cutter, and Canner. I live in the top two. Rarely do I choose Select, but it's important to know and understand the function of each grade.

Prime graded beef is the golden ticket. It's the highest grade in the system, signifying that its juiciness, tenderness, and flavor are absolutely top-notch. Prime means that the beef is abundant in intramuscular fat—which gives your dish that mouthwatering zing—its rib eye muscle is moderately firm, and its meat is light red in color and fine in texture. I'm getting hungry just thinking about it. Young, grain-finished steer and heifers are the ones that usually populate this premium grade. Bullocks and bulls can also be graded Prime, but their beef isn't as palatable as that of steer and heifers, and cow beef is not even eligible to be in this category. Breed-wise, Black Angus is the main player here, due to its consistency when it comes to marbling and yield. All of these factors combined whittle Prime's pool down to less than 5 percent of the entire country's beef population, which bumps its price up and is the reason it is generally so sought after by restaurants and select retail establishments. This is my go-to when I want to cook up a thick, juicy, all-American–style steak.

Choice is also considered high-quality beef but has a little less marbling than Prime. However, its amount of intramuscular fat still delivers delicious-tasting meat, and is a little easier on your wallet, making it a popular grade, widely available at restaurants and supermarkets. Cow beef, which can only be graded as high as Choice, falls into this category together with the rest of the cattle that meet the necessary requirements when ribbed, such as having a small amount of marbling, a slightly soft rib eye muscle, and meat that is moderately light red in color and fine in texture. If you don't have access to Prime, then Choice is well worth it.

Select is leaner than Prime and Choice, which causes it to be less juicy and flavorful due to such slight marbling, but it is still fairly tender. Widely available in the retail market, this grade of meat, which also must be thirty months of age or less, has a rib eye muscle that is moderately soft and meat that is slightly light red in color and fine in texture. Note how as the grade quality decreases, the meat color gradually becomes darker. It's something to keep an eye out for when choosing your cuts.

Standard and Commercial graded beef are practically devoid of marbling, which makes them considerably less flavorful and tender and a little chewier than the first three grades. However, when necessity trumps high-quality grades, these economical cuts are the way to go. Cattle that are too old for the first three grades fall into these categories. The rib eye muscle is soft and the meat is slightly to moderately dark red in color and starts to feel somewhat coarse in texture. Say you have a grass-fed animal that's just been used for milking. It's eight years old. There's no way it has intramuscular fat if all it has eaten throughout its life is grass. In order to remedy this before it goes to market, the grower will likely transition it into a grain-fed diet for about one hundred days in hopes of injecting it with a little bit of flavor. It won't fall into any type of high-grade beef, but there's definitely a market for cheap beef that still has some quality to it. In retail, you will be able to identify this beef because they are the cuts that are ungraded. They do not receive the USDA shield and they're usually sold as store-brand beef with lower price tags.

Utility, Cutter, and Canner are the last three categories, and they include cattle of all ages (but mainly those older than thirty months) that did not meet the requirements to fit into any of the higher USDA beef grades, such as milk cows that are no longer in service and some downers. This meat can range in color from slightly dark to very dark red and is usually coarse in texture and devoid of marbling. It's really not something you will need to learn to identify because beef that falls into these categories is seldom sold at retail. This less expensive product is usually used in pet food, processed foods, and canned goods, and is shipped to prison systems and to the armed forces. Military MRE (Meal, Ready-to-Eat) boxes include pouches filled with three thousand calories worth of the nutrition a soldier needs to have enough energy for the entire day. Why use Prime or Choice beef here, if you can't even taste it? The goal is to get protein into the diet, and from a nutritional standpoint, Utility beef has the same value as any other beef grade.

Like Standard and Commercial, these last three types of beef also remain ungraded and are deemed as "no-roll" product. As we've discussed earlier, when a USDA grader at a harvesting facility has made their decision regarding the quality of the beef at hand, they roll a stamp down the length of the animal to clearly indicate its grade on each future primal cut. If they don't grade it at all, that means it wasn't rolled or stamped. Animals that were raised and finished per higher-quality standards but are found to have bruises or any other type of visible mistreatment will not get graded either. That's why I keep insisting that it's in the grower's and harvesting facility's best interest to feed the animal right and treat it humanely throughout its life.

As I've mentioned before, the USDA grading system is voluntary. Growers and harvesters don't have to be graded in order to sell their product. They can have no-roll product all they want, and there's a market for that. However, at some point, a savvy consumer will realize, "Hey, this beef isn't all that great." Sure, it may be the least expensive, and readily available at big-box supermarkets, but if you look at the label, there will be no mention of Prime, Choice, or Select. With no-roll product, the quality may suffer, but that doesn't mean it's bad. It will really just come down to what you're looking for.

My recommendations? If you're interested in having a great steak, then yes, go for the Prime, Choice, or Select beef. If you need to feed a big family, or as many troops as possible in the army, then quantity and cost trump quality, hands down.

When beef is rolled, that grade stamp information must be added to a label to ensure that it reaches the end consumer. That means that when I receive the vacuum-packed primal cuts from the harvesting facility, they must be clearly labeled with their corresponding grade and any other pertinent specifications related to breed type or raising claims, such as all-natural, and feed type, like grain-finished or grass-fed, for instance. Although grass-fed is well regarded by many consumers in the retail sector, it usually doesn't grade too high given its lack of marbling. No grains, no gain in intramuscular fat. Another spec that is certified and added to the label refers to religious attributes, such as kosher beef.

Traditional Jewish law not only doesn't allow kosher cattle to be stunned before they're slaughtered; it also requires that the animals be harvested and processed at a kosher-certified facility. Additionally, only the forequarters, or the front end of the carcass, may be used for kosher beef. Therefore, any primals coming from the rear end, due to religious laws, are absolutely forbidden. That automatically means that there's no such thing as kosher filet mignon because that cut comes from the animal's midbody to hind leg section. To satisfy kosher consumers' desire for this cut, a rib eye is cut in half lengthwise to make it look like a filet mignon, but it's clearly coming from a different primal part. Since only 50 percent of the animal can be used for kosher beef, it's more expensive regardless of its quality—the rest of the animal is sold at a discount.

Once the primals are packaged, labeled, and shipped from the harvesting facility, their next stop, in my case, is my warehouse. That's where, together with a USDA inspector at my facility, we ensure that, once we cut and package the beef for my clients, the grade and any other certifiable and vital information is carried over to our own labels. Labeling is essential at this point in the process because by then stamps are usually no longer visible. The Food Safety Inspection Service oversees label development and compliance and enforces beef labeling laws. Falsely labeling products at any stage of the process can be punished with fines of up to $10,000, warnings, and up to three years imprisonment.

Back when my dad started working with my grandfather, these laws were yet to be truly enforced. All the meat got supplied to the Fourteenth Street Meat Market by train along what is now known as the High Line. If those rails could talk. From the mid-1800s through the 1930s the High Line train actually ran on street-level tracks along Tenth Avenue, aka Death Avenue. Precarious conditions for pedestrians led to numerous accidents, and as train-related deaths soared into the hundreds, the city's Transit Commission had to step in with a plan to elevate the tracks. By

1933 the first trains carrying throngs of meat, dairy, and produce began running along what was then known as the West Side Elevated Line.[5]

When my dad was old enough to start helping my grandfather out at the shop in the 1950s, they had no choice but to procure meat from the Fourteenth Street Meat Market because all trains with food led there. Labor unions at the time reigned supreme and strikes abounded, which forced butcher shop owners and employees to figure out other ways to get their needed supply of meat. Train conductors coming into the city from the Midwest were clueless about New York City's issues until they rode into what is now Chelsea and, a few blocks short of their Fourteenth Street destination, suddenly got held at gunpoint. They couldn't have imagined that city dwellers would behave like Wild West cowboys. While the gunslinger forced the conductor to stop the train, a group of guys would climb onto the cars, grab the heavy slabs of meat, and hurl them over the rails and onto the pavement. Waiting about thirty-five feet below the tracks were the luggers, who caught the flying meat, loaded it straight onto their trucks, and sped off.

At one point about 30 percent of all meat in New York City was from the black market, until the FBI cracked down on this racket and made it a criminal offense. My dad was one of those luggers lying in wait; it was just the way it was done back then to keep your business running. He remembers those days vividly—the tension, the pressure, the urgency—and that's why he refuses to set foot on the beautiful tree-lined High Line today. "That's where the meat was," he grunts. It's hard for him to ignore those memories when looking up at those abandoned and now refurbished rails. By the 1960s and '70s, trains grew scarce as trucks began to play a larger role in the food transportation industry, and by the 1980s they came to a complete stop.

Safety, labeling, and traceability of the product didn't carry the same importance in those days as they do now. The meat that was flung over the rails and hit the street was automatically smeared with tar, but all that mattered then was cracking the system and getting the product needed to survive in that mafia-run industry. Traceability in the beef industry became a front-and-center issue a few decades later, when the first cases of bovine spongiform encephalopathy (BSE, aka mad cow disease) spread from cattle to humans in the mid-1990s and wreaked havoc on the industry in the early 2000s. BSE is a progressive neurological disease that damages the animal's central nervous system from the brain all the way down its spinal cord. It was first identified in the United Kingdom in 1986. It may have originated as a result of feeding cattle meat and bonemeal that contained BSE-infected product. The outbreak among cattle peaked in the United Kingdom in 1993, and the first BSE variant, called Creutzfeldt-Jakob disease (vCJD), was detected in a human and linked to mad cow disease in 1996.

Things got bad for us here in the United States in December 2003, when the USDA announced the first case of suspected BSE at a plant and the following day recalled beef from cattle killed in that plant. The world cut us off. This brought the US beef export industry to its knees, not only due to the outbreak but because it shed light on the fact that beef exported from the United States, which other countries paid a premium for, might have actually originated in Canada or Mexico. It was like the olive oil debacle when consumers realized misleading labeling made us think we were buying olive oil from Italy when in fact it was produced with olives harvested in other countries. So how did we get here? Simple: the North American Free Trade Agreement (NAFTA). When it was enacted in 1994, this agreement established a free-trade zone among Canada, the United States, and Mexico. Up until then, if a label mentioned a product was from Canada or Mexico instead of the United States, many American consumers chose the US-labeled product. Canada and Mexico didn't like this because it placed them at a disadvantage. So, when the NAFTA deal was drawn up, it was decided that labeling the country of origin among these three countries would no longer be necessary. Congress actually did pass the Country of Origin Labeling (COOL) law in 2002, but it took years for the USDA to really enforce it. Hence the conundrum we faced when BSE arose in 2003.

Eventually, using the original infected cow's ear tags and genetic testing, our government was able to trace its origins back to Alberta, Canada, noting that it had been imported in 2001 but was born in 1997, only four months shy of when an American and Canadian ban of the use of brain and spinal cord tissue in cattle feed had gone into effect. But by the time this news became public, the harm had already been done. Around thirty countries decided to stop buying US beef, including Canada, which imposed a partial ban. When the cow's Canadian origin came to light, the tide turned and the focus was now on our northern neighbor, but there was still a lot of work to be done on our front to get our international buyers to once again trust our product. With time, we were able to recover our exportation licenses, but in early 2006, Japan reimposed the two-year-old ban, which they had lifted just a month earlier, when they received a shipment from New York–based company Atlantic Veal & Lamb with veal that still had their vertebral column. By then, this was considered specified risk material and was required to be removed to further avoid possible BSE risk for animals over thirty months of age, though in the United States it isn't considered at-risk material when it comes from cattle younger than thirty months. What created the alleged misunderstanding was that the deal with Japan was to remove the vertebral column regardless of the animal's age.

Now, cattle older than thirty months, in addition to having all SRM removed—including bones along the spine, which eliminates porterhouse and T-bone steak cuts from this older group—must be processed separately from younger cattle. However, most harvesting facilities insist on removing all bones from these older animals out of an abundance of caution. A BSE case can really wreck your business, so why risk it? No mad cow disease has been found in animals younger than thirty months, so I don't purchase any animals older than that for our facility. We've completely cut ourselves out of that category. That's part of our program rules: no animals over thirty months of age, period.

Atlantic Veal made national headlines—I'll never forget this because we bought our veal from them at the time—and people were furious because it dealt a huge blow to our industry, especially given that the trade relationship with Japan signified millions of dollars for the US beef industry. The problem was eventually sorted out and the cattle trade returned to normal. Meanwhile, the industry continues to work on firming up its domestic traceability system. Since we are acutely aware that there is still room for improvement, we already have pilot programs in place in the hopes of eventually expanding superior cattle traceability to a national level.

USDA guidelines really push us to excel when it comes to quality, making international buyers hungry for our high-quality, grain-finished meat, but we also import beef from other countries. According to the USDA, we import more cattle than we export, mainly from Canada and Mexico, with the latter providing almost 60 percent of imports.[6] These cattle from Mexico are mostly lighter in weight, so they're shipped and finished in US feed yards and then harvested in one of our domestic facilities. That means that close to 10 percent of the cattle on US feed yards weren't born here. So, is our beef American or Mexican?

Furthermore, about 75 percent of grass-fed beef sold in the United States is actually raised abroad. Most consumers who buy grass-fed beef thinking it's better for the environment are not aware that there is a 75 percent chance they're buying something that was born and raised elsewhere—likely in Australia, New Zealand, and parts of South America—and then shipped here to be processed by a US harvesting facility. I don't think anyone has taken the time to add the greenhouse gas emissions from transport to the carbon footprint calculations for grass-fed beef, but they really should.

You may be thinking that as a careful consumer who reads the labels, you're safe because you always choose beef that is labeled "product of USA." Unfortunately, that information can be misleading too. If meat is processed at a USDA-inspected facility (which is required for all imported beef), then by law it can be labeled as a US product. But didn't the whole BSE calamity give way to the enforcement of the COOL Act? Yes, for a while, and then in 2016 it was repealed for beef and pork. Therefore, beef and pork that originate in other countries but are harvested here do not need to reveal their country of origin on the final consumer label. This is not only a disservice to consumers but also to our domestic growers, many of whom are independent farmers who don't have the means to go up against an influx of imported beef with misleading labels.

I understand the need to import beef to keep our supply robust, but I'm also a staunch supporter of our domestic beef and believe that consumers have the right to know where their product originated so that they can make their own decisions when purchasing beef. As the government and industry figure all this out, I say we change the current "raised and grazed in the United States" motto to: *Born, raised, grazed, and processed only in the USA.*

# PART TWO

# THROUGH THE MEAT PURVEYOR

# 5

# A BUTCHER'S ROAD TO MEAT PURVEYORDOM

All it took was one fight. One trivial childhood fight at the turn of the twentieth century is what ultimately defined my family's history and business and the course of my own life. A fistfight, on a cobblestoned street in Naples, Italy, as luck would have it, right outside a butcher's shop. The young boy receiving angry punches and defending himself with all his might, only to get knocked down by one final blow to the face, was my great-grandfather Anthony LaFrieda. Defeated and alone, he hobbled over to the curb, sat down, and let the tears of frustration and humiliation streak down his flushed cheeks.

From his doorway, the local butcher watched the scene unfold and took pity on the beat-up kid, so he told the boy, "Come into the shop, let me put a steak on your eye and we'll get the swelling down." Grateful, my great-grandfather followed the man into his store and watched as he grabbed a slab of meat from the counter, walked over, and handed it to him. Sure, he'd taken a beating, but those fights were commonplace back then, and this kid still had all his bones intact. And while losing a fight is embarrassing, it's not the end of the world, which is why the butcher was baffled by Anthony's downcast demeanor. After a brief pause, my great-grandfather finally looked up to the butcher and said, "My father is going to kill me because I got into a fight." His dad was extremely strict, and one of the household rules he drilled into his children was that fighting was absolutely forbidden. Anthony knew that his soon-to-be black eye would be a dead giveaway, so he would have no choice but to tell the truth and face the daunting consequences. No wonder he was so distraught. The butcher listened attentively, and then said, "Tell him the truth. And tell your father you may have lost the fight but you got yourself a job." Anthony's jaw dropped. Did this man just say he was offering this bruised kid work? "You come here, and I'll teach you how to be a butcher so you can learn the trade," added the kind man.

A job at a butcher's shop was a big break in those days when opportunities were scarce. Grateful for such a stroke of luck, Anthony eagerly accepted and, unbeknownst to him, took the first step into a business that would become a huge part of my family's legacy. As Anthony learned the tricks of the trade, what started as a job quickly became a passion. The butcher saw his raw talent and mentored him into adulthood. In 1909, married and already with a few children in tow, Anthony decided to take his growing family and his butcher's skills and join the millions of Italians emigrating to the United States. That's how they came to disembark on Ellis Island, with his trade under his belt, his family, and a dream. And that's how my grandfather, Patrick LaFrieda—the first Pat in the family—became part of the first generation of American LaFriedas, born in New York in 1913, one of five sons and two daughters.

By 1922, Anthony had managed to collect enough savings and local experience to open his own retail butcher shop in Flatbush,

Brooklyn. To run the place, he called on the help of his five sons, who with that on-the-job training all went on to become butchers in their own right, working for other companies or eventually opening their own stores. That Flatbush shop was my grandfather's school; that's where he was introduced to butchery, and he never looked back.

The art of butchery takes fine-tuned skills and great physical endurance. This craft is not for the faint of heart. A butcher must first understand the anatomy of the animal in front of them in order to learn how to handle the primals. The different cuts of meat should be drilled into your brain to the point where you can recognize them in your sleep, so that when your customer's specs hit your table, no time is wasted and you get straight to work. You must know how to expertly wield a knife and execute precise cuts, slicing the meat before you with absolute determination to produce a smooth surface for the eventual cook. Sharp knives are crucial in the process. When a knife is dull, it not only slows down the production line but also causes the dreaded staircase effect on the meat, which is a huge rookie mistake. Beef must be cleanly and swiftly sliced through, never sawed down.

The job itself is fast-paced, grinding, oftentimes exhausting, and extreme attention is required at all times. Nicking yourself is a given, but losing your focus could mean losing a finger. One small distraction, one unnecessary glance up from your knife to the clock or your colleague and you could be toast, especially when it comes to the most revered and panic-inducing machine of them all: the bandsaw. This is the most dangerous tool in the building; it can cut through bone in milliseconds. If you handle it with fear and doubt, if you get distracted, if you don't apply laser focus on pulling the meat through the saw rather than pushing it through, it could leave you maimed. That's why my dad proudly says, "I've been working on a bandsaw since I was twelve years old, and I still got ten fingers. A lot of my competitors can't say the same."

It takes a special breed of people to muster the strength, stamina, and stomach to endure being on their feet, all night long, cutting meat with sharp tools, in a 35-degree room. That's why hiring new butchers is far from easy. Aside from first testing their knife skills and seeing if they can follow the crucial safety measures on the bandsaw, I also need people who are malleable, people I can train to do things our way. The only road to becoming a butcher is to be in an environment where you can practice every day, cutting meat until you perfect your skills. We made the butchers we have because we need them to come through for us. If one butcher of the twenty or so working a shift is unable to follow a client's instructions and doesn't cut the meat per the restaurant's specs, that customer may very well not just send the meat back to us the next day; we could lose them for good. We don't just bake cakes at LaFrieda Meats. Every night we process about eighty thousand pounds of fresh meat by hand—the way my grandfather used to do it—for around six hundred restaurants.

When I walk into the production room to join the butchers on the floor, I come in carrying one knife, a twelve-inch-long cimeter. That's it. First off, its upturned curvature means that its point will never puncture or poke anything I'm cutting. When new, it's nice and thick so it has the backbone to cut through fat and steak. With use, the blade becomes narrower, and that's when it's best to use it to sinew off meat and slice things really thin. Because of my six-foot-three-inch height, regular tables are a little too low for me—unlike when I first started out as a ten-year-old kid standing on a milk crate—so, instead of hunching over and hurting my back, I use this long knife. I've always done it that way, and I use it for everything, something the newer guys usually don't understand. One night, I could tell these fairly new guys were shooting side glances my way as I headed over with my trusty cimeter to a table piled with chickens that needed to be cut. They'd never seen me cut meat before, and I assume they thought I didn't know what I was doing because they kept trying to tell me I was using the wrong knife for chickens. I looked them straight in the eye and asked them to reevaluate me in ten minutes. They stood by, quietly observing as I bypassed their cleavers and went straight to work. I finished that entire order for them in ten minutes and they had to eat their words. For the rest of the night, they followed my lead and, together, we managed to finish the remaining orders three hours earlier than usual, which allowed them to move on to the next tables piled with fresh meat lying in wait. Of all the cutting tools that exist, a cimeter and a boning knife are all a butcher really needs.

One of our longest-running and best butchers is a guy who was hanging out outside our Leroy Street building and was brought to my attention right around when Shake Shack reached out to us to start hand-forming their burgers, which was something we didn't do at the time. When someone mentioned that the guy was clean and in desperate need of a job, I went outside to meet him. We started chatting—by means of a translator because he didn't speak a lick of English—and I came to find out Eduardo Martillo had actually been working at another shop forming four-ounce patties, so I hired him on the spot. Soon after, I was recovering from my first hernia surgery but had a leg of veal I had to break down for a client. The new guy quietly observed the situation and then quickly sprang into action and signaled what I understood as "Get out of the way, I'll do it for you." As I watched him section out the leg, I said, baffled, "Why do we have you making burgers?" He's one of our most skilled butchers and is still with us today.

Butchers can be a rowdy crowd, so you have to know how to hold your own in their presence. I can still joke around with my old-timers, the guys I've been working with side by side in our production trenches for the last fifteen to twenty years, but when I set foot on the floor to cut with my butchers, usually they immediately grow quiet with respect. It comes with the territory of being the boss. Do they want to work next to me? No. They all know I have one speed, and that means it's not going to be a relaxing night. Which is fine, I don't blame them. But when they're in the weeds, they know I'll get them through the night. I like to cut with them. I want to cut with them.

Cutting meat is therapeutic for me. It's like how I imagine others might feel carving something from a piece of wood. When I'm having a stressful day, there's really nothing like stepping up to a table and losing myself in the cutting process for a while. I come out of there renewed and ready to face whatever's waiting for me. To be honest, I also thrive on the adrenaline rush of being on the line and having to meet our strict deadlines. For example, the Boston and DC trucks have to leave our facility by 1:30 a.m. to make on-time deliveries in those respective cities, which means that these orders of meat need to be cut by 11:30 p.m. to allow enough time to complete the invoices, pack, and load the orders onto the vehicles. That's when I turn up the pressure on the floor.

Someone has to be the asshole in business. And sometimes that responsibility falls on me. It's not just about training the butchers and making sure they're up to the challenge, it's also about dealing with all of those different strong personalities trying to coexist under one roof, my roof. My goal is to strike a balance so that everyone gets along and we can ultimately be one big family. It takes a great deal of time to bring a team like that together, and sometimes it doesn't last long. We're talking about a crew with super tough skin, butchers with sharp tools, knives that have easily gone from cutting a slab of meat to being pressed against another guy's neck. That particular scene happened recently, and I had to jump in and stop this big guy, a former baseball player, from taking it too far. He was immediately let go. No one can behave like that at work, no matter what beef they have with each other, so to speak. Everyone has the right to a bad day, and when I see that one of my butchers is being pushed past his limits and might be on the verge of cracking, I send that person home to sleep it off and come back the next day with a better attitude. Thankfully, I have such a great team that those confrontations aren't the norm. What I aim to foster is that unique sense of humor and camaraderie that can bloom on the line. When that magic chemistry happens, I know the night is going to flow smoothly.

So how did we get here? How did my family go from my great-grandfather's retail butcher shop in Brooklyn to fulfilling six-hundred-plus restaurant orders from a 35,000-square-foot facility in New Jersey? It all began with the great strike wave of the late 1940s, the largest strike wave in US history. Workers in industries across the board participated, from electrical and steel workers to coal miners and railroad engineers, as well as more than one hundred thousand meatpacking employees. It was such a massive strike it practically paralyzed each and every one of those sectors.

In itself, it's very difficult for a retail butcher shop to make a profit; add an almost monthlong strike to the mix, and the other pockets of strikes that ensued after that, and what they had upon them was a perfect storm, which led to many shops having to close their doors for good. Amid the mayhem of those years, my grandfather Patrick and his brother Lou's survival instincts kicked into high gear and, much like Al Capone running alcohol during Prohibition, they realized that the chaos created an invaluable opportunity. Restaurants were paralyzed by the ongoing dwindling meat supply, so Patrick and Lou decided to drive out to local farms in New Jersey and upstate New York to help fill that void. They weren't union members, so nothing prevented them from selling meat straight to restaurants, but they preferred to keep the unions in the dark to avoid having to face their wrath. And it worked. By delivering cuts to this captive audience of restaurants in need, they managed to get their foot in the wholesale market door. Always ahead of his time, my grandfather dipped into the farm-to-table trend decades before it became a thing.

This successful shot in the dark eventually allowed my grandfather Patrick and great-uncle Lou to venture out on their own with their new roster of restaurant clients. And in 1950, they opened their very own shop on West Fourteenth Street, right in the heart of the Fourteenth Street Meat Market: the original LaFrieda Meats. Once the strike wave rolled by and the economy and industry began to rebound, they went back to procuring their meat straight from the market—you didn't want to ruffle any mobster feathers by doing otherwise—but their business model had evolved, and they had made the family's first official move from retail butchery to meat purveyordom.

As my grandfather and Lou picked up more restaurant accounts, they were eventually able to move a couple of blocks south to a second-floor space in a dilapidated old tenement building on Little West Twelfth Street. That's around the time my dad started working at the shop as a teenager, pretty much against his will. He not only had to hustle at the Meat Market, competing with the bigwigs to get his hands on some decent meat, he then had to haul the two-hundred-pound hind saddles of beef over his shoulder, up the flight of stairs to their shop, and stand on his aching feet for hours on end to break the cuts down, while he dreamed of cars or playing music. As if that weren't enough, since the neighborhood was always in such disarray, when everyone went home and he was left behind cutting meat and wrapping up his day's work, his father would lock him in the room with the meat on hooks because he didn't want any of it to get stolen. One of the reasons beef hung on hooks back in the day was basically the rodents roaming those grimy streets. Mice won't eat meat, but rats sure will. It was bad.

When I think of those dire conditions my dad endured as he came up in the trade, I understand why he so adamantly tried to push me far, far away from the family business. It's back-breaking work in freezing conditions with grueling hours, and not everyone can handle it, but he displayed the talent and eventually became an expert in his own right. I will never stop admiring his expertise and leadership skills, a true testament to decades of tireless work.

By 1964, my grandfather and father—who was only eighteen—took over LaFrieda Meats from my great-uncle Lou and renamed it Pat LaFrieda Meat Purveyors, continuing the line of business the two brothers had inaugurated more than a decade earlier as wholesale butchers, which meant supplying meat to restaurants rather than the general public. My father always preferred it that way; he just found retail butchery wasteful. The one time he did give it a shot, it only helped prove his point. After enduring a few days with small retail orders, the last straw was when a customer came in and ordered chicken wings. He fulfilled the order, then turned around to his dad and said, "Now I have a chicken with no wings, what am I gonna do with it?" Having to interrupt cutting meat for larger restaurant orders for a few measly chicken wings also bothered him, so that was that. It wasn't until I stepped in that we reinstated a retail sector, but that's a story for another day.

**BEEF HAS ALWAYS BEEN IMPORTANT** in the meat industry, but the problem back in the 1980s was that we just didn't sell enough of it yet. We moved more veal and lamb than beef—compared to other companies—in part because, for quite a while, there were insurmountable obstacles in our way. First off, for years on end the Meat Market dictated policy and suffered no repercussions or consequences because the mafia controlled the beef and poultry in the city. So much so that in 1986 congressional hearings addressing the mob, it came to light that chicken mogul Frank Perdue had actually met with renowned New York mafia boss Paul Castellano Sr. about his business. Only when Big Paul offered to buy Perdue's product through the family's distributing company, Dial Poultry, was Perdue able to get his brand in the city. That's how far the mafia's power reached in the business at the time.

Thankfully we had no grievances with the mafia in this respect.

My grandfather was friends with Paul Castellano Sr. Our families have been friends for a century and we have a long history of working amicably together. That was the key to handling those days: remain at a friendly distance because even if you made nice with one family, there were still others to deal with. So, did we have to kick money up around the holidays? Yeah, sure. If you didn't, you ran the risk of falling out of the mob's good graces. But we never had to pay weekly protection money like most other businesses in the neighborhood. And that was mostly because my dad bought meat from their distribution houses, some of which were owned by the Italian mafia, others by the Jewish mob—there was corruption everywhere you turned; it was impossible to avoid it. But one thing's for sure, as good businessmen, they weren't about to come down on their own customers. By purchasing product from those houses, we avoided the risk of imposed strikes or protection payouts, which was amazing because, back then, when you were "required" to make a payment, you either paid up or ended up in the river.

At the end of the day, since none of them wanted wars, territories were marked and peace was upheld. Once we outlived the mafia, we were finally able to cut the middleman and start dictating our own policies. That's when we began to grow, and a few years later, we were at last able to order a trailer-load of beef directly from a harvesting facility, and never looked back.

---

**NOW, AS MEAT PURVEYORS,** one of our main roles is to get the pulse of the market and what our restaurant clients need to fulfill their end customers' demands. In a day, I get a pretty constant stream of texts and on average receive calls from thirty to sixty chefs. Some calls are during the day, others are after midnight. The conversations range from, "Hey Pat, what kind of meat should I use for my steak frites at this price point?" to figuring out if it's possible to provide them with a specific cut, down to the bones and ounces to satisfy their needs. Like this one call I got recently that left me slightly puzzled. When I picked up, the chef went straight to the point.

"Pat, is it possible to cut me a double-bone queen steak?"

I honestly didn't know what he was talking about, so I asked him to send me a photo. From what I could see, a double-bone queen steak was basically a rib steak with two bones attached, like a rib roast, cut for two, and, according to this chef, it had to weigh 38 ounces. However, a single bone alone weighs 38 ounces, so if there were two bones, that would push the cut close to 80 ounces. That's when I looked at the photo again and noticed there was no rib cap meat on that steak.

"OK," I said, "you buy rib cap beef at twenty-six dollars a pound. I could take that off, and you would have what you're calling the 'queen' cut, but you're not going to have the spinalis muscle." That was fine by this chef. So I managed to decode his request and figured out a way to make it happen for him. I had to take apart a rib roast, remove the spinalis muscle, and trim it, roll it, and cut it into portions. Then I had to take the rest of it and cut it at every two bones, even though there are seven bones on the primal cut. So how do you get four two-bone rib roasts if there are only seven bones? I live for these puzzles and challenges. If I only gave the chef three versus four, it would cost the restaurant ten to fifteen extra dollars a pound. It had to be four. So I grabbed a center bone and split it right down the middle with a bandsaw.

Even though it was technically a split bone, it appeared to be the same size as the rest, so for that chef's purposes, there were now two bones and the puzzle had been solved—plus, we saved our client money and made sure to use that primal cut efficiently. From my perspective, that's a job well done on all fronts. That's what it's all about.

6

# AN EDUCATION AND A DREAM

I quit finance to join the family business, against my dad's wishes. If you've made it this far, you know that it all began on that special day when my father brought me to the shop on Bleecker Street when I was ten years old: that's when I knew I wanted to be a butcher. So how did I get from there to here? Let's go back to those formative years when my dad allowed me to work with him in hopes that I would see firsthand how strenuous a job it was and fall out of love with it. What he didn't realize was that those early days really turned out to be the best basic training of my life.

My first butcher shop memories are of my father and grandfather's Bleecker Street building. Their move to Bleecker from Little West Twelfth put them twelve blocks south of all the Meat Market madness, situating them in a sweet spot that offered a respite from the chaos but was still within reach for business purposes. My dad can't believe how clearly I can still picture that shop. As you walked through the inconspicuous doorway, you'd step on sawdust on the floor, which was used for cleaning—from an inspection perspective that would not fly at all today!—and find the production area stationed at right. That's where I stood at ten with my dad and grandfather watching over my shoulder as I split those top rounds of beef and tied them with a butcher knot—a recently acquired skill that made my heart swell with pride—and made roast beefs like my life depended on it.

Grandfather Pat and my dad weren't big on talking while they were cutting meat in that small room, where we were all pretty much shoulder to shoulder working on the same six-foot-long table. Their calm filled me with even more respect for them, my role models. By then, my grandfather was older and overweight and would sometimes have to lean on the table to hold himself up while cutting beef, but nothing could keep him from doing what he loved. He was a gentle and kind man, and so loving. He always covered his grandchildren with hugs and kisses each time he saw us. I still remember the scratchiness of his scruffy beard against my face, but I never minded it. It was always wonderful to give my grandfather a hug. As kids, he'd take my siblings and me to Howard Johnson's for breakfast on the weekend and buy us Chuckles candy. I loved him so much.

I also loved the relationship he had with my dad. They had this great, funny, back-and-forth banter that reminds me of the relationship my dad and I have now. My dad would make fun of my grandfather because he kept crashing his car. It was quite the sight. Since my grandfather was really overweight, he'd use his cane to reach the brake pedal, which many times did not work out as planned, ending in accident after accident. It happened so often, my father nicknamed him Crash Gordon. Another one of his quirks was that sometimes he'd just go missing. My dad would have to search the neighborhood for him, and nine times out of ten, he'd find him at a nearby hot dog cart covertly chomping down on a bunch of hot dogs while expertly avoiding our family's judgment and reprimand because he was supposed to be on a diet. When my dad finally spotted him down the block, the shout fest would begin, and I couldn't help but chuckle at my grandfather's sweet, mischievous ways.

A few steps past the production area was my dad's desk. The shop was really just one open floor with a refrigerated room, the freezer, the processing room, the "office," which was a desk situated in the middle of the room, an inspector's office, and farther back, the little restroom. That's where I came across my first *Playboy* magazines, which were stacked in a corner, and the first time I ever saw a naked woman. One of the employees, Pino, would rip out pages of his favorite models and pin them to the wall—I know, unheard of nowadays, but that was a normal thing in the 1980s, and no one really gave it a second thought. For me, at that age, when my eyes wandered to these photos or caught a glimpse of the magazine covers, I'd quickly glance away, embarrassed, as if my parents had been keeping watch over my shoulder. Those magazines didn't make it to our next location, but I'll never forget that moment of prepubescent innocence.

What used to be our Bleecker shop became the Village Apothecary when we moved out in the early 1980s, and is still there to this day. The reason we made that next change in location was that my father and his sister, my aunt Lisa, who worked as the business's bookkeeper, had managed to save enough money to purchase a block-long property on Leroy Street, which years later would be renamed Pat LaFrieda Lane in our honor. This is where my real training began.

Buying that property was a genius move on my father and aunt's part, especially at a time when everyone else was leasing.

It was that Italian mentality cemented in Brooklyn, which was all about buying two- or three-family homes to subsidize mortgages and taxes. My dad repurposed this tried-and-true method and applied it to the business. They bought the entire block, kept the space needed for our business, and rented out the rest of the spaces to other companies, including a transportation business, a mechanic, and an artist. At first, we just occupied the property's Leroy and Washington Street corner. During the winter of 1981–82, I'd accompany my dad to that corner property and watch him in awe as he slowly built that place up himself—he didn't have enough money to have anyone else do it. I remember the fire in the middle of that big, open warehouse burning through the day to keep him warm. And he did it; he made it happen. My admiration and respect for him had no limits.

After moving into that corner building, as the years went by and tenants moved out, we slowly began to take up the extra space to accommodate our growth. By the end of our stay there, before we moved to New Jersey, all that was left was LaFrieda Meats and an art gallery. I even lived there for a while, in the second-floor apartment located over the space rented by the artist at 620 Greenwich, until 1999, when I bought my first house in Bergen County, New Jersey. So yes, my dad and aunt's multiuse purchase was absolute genius, and served its purpose ten times over.

Bearing witness to my dad's sacrifices, watching him work night after night at the table, made an indelible impression on me. As did my mom's unconditional love for my siblings and me. She had been a nurse for many years, but after having four children, she turned her attention to just us. If she had to get up and make four different breakfasts to satisfy each of her kids' needs, then she'd do it. She also kept active working in real estate, and when

Aunt Lisa retired, she stepped in as our bookkeeper. My mom still works at our New Jersey facility once a week taking care of restaurant bookkeeping.

There was nothing my parents wouldn't do for us, and as the eldest of the four kids, I quickly took notice. All of their hard work and sacrifice made me incredibly aware of the expenses we created. I remember once, when I needed a new pair of sneakers, I didn't want to ask for them because I knew all the work that it would take to generate that money. It gave me a deep respect for everything they did, like putting me through an expensive high school with no loans or financial aid. My dad would often kid around and say, "Hey, I could be on a beach right now," but instead he and my mom were devoted to our upbringing and working relentlessly to provide us with only the best.

I have fond childhood memories of both my parents, but I'd be lying if I didn't say that as a kid my dad was God to me. Those first few years when he started taking me to the shop in the early 1980s were also the first time I saw my dad being treated differently. There was a palpable respect that wasn't present during interactions with people who didn't know what he did or who he was. But over on Leroy Street, he was the owner and the boss, and everyone looked up to him. My grandfather and father were also very well respected in the industry in general because they had earned a reputation of being men of their word. Theirs was not like other meat companies that built a business and sold it off, or even worse, owed money to the Meat Market, claimed bankruptcy to avoid paying off their debts, and then reopened under another name. Everyone in the industry knew that my grandfather and father invested their lives in their business and that they were good for it, always paid on time, and weren't planning to go anywhere anytime soon. They were in this for life. All of that went a long way among their peers, and it also created an even greater level of respect among their employees.

The way I felt as a kid, at work with my dad and my grandfather, the way everyone respected them, the way I admired them—when I tried to express this, well, let's just say it didn't translate well with the affluent kids back at my prep school in Brooklyn. But I didn't care what they had to say. My dad sacrificed everything to send me to that private school to give me a solid education and get me off the streets, which back then were a breeding ground for the mob. It was an odd upbringing, seesawing back and forth between my preppy classmates and the gang members in my neighborhood. It definitely took a little while to adjust. The first few times I went home wearing my school uniform—a sports jacket and tie—I got these odd looks and was challenged by some local kids, so I quickly learned to take the jacket and tie off before exiting the bus in Bensonhurst, fold them over my arm, and inconspicuously carry them back down the blocks that led me home. Little did I know that this experience of living in two contrasting worlds was preparing me to navigate, many years later, between my suit-and-tie meetings with prospective clients and my fleece-covered nights cutting meat with my butchers.

When I wasn't in school, I was working with my dad, who was trying to keep me far away from the dangerous element brewing in our neighborhood. That included weekends and any days off from school. I guess he figured he could kill two birds with one stone: keep me out of trouble while also showing me how strenuous it was to be a butcher. He'd wake me up in the middle of the night so we could make it to Leroy Street before dawn. We'd grab some fresh bread or bagels from the bakers in the area—the only other people up and working at those hours—and then head into the building to start the shift. And no, I wasn't cutting meat all day. His philosophy was that if I wanted to learn the business, then I had to learn every single grueling aspect of it, including carrying twenty-five-pound primals across the room several times over, sweeping the sawdust off the floor, and even cleaning the fat trap—not exactly the definition of fun, but I still loved every minute of it. I felt useful and I was learning so much that the next shift couldn't come soon enough.

Aunt Lisa would come in at around nine in the morning and overlap with us for a few hours until it was time for us to go home. Then she'd stay a while longer to take care of the bookkeeping and take down the afternoon orders. My mom always lent a hand when they needed extra help, until she started working there full-time once my aunt retired in 1994. It was the epitome of a family business.

From the start, I was always the lowest man (well, kid really) on the totem pole. My dad never wanted his crew to feel as if I was above them. And I knew better than to act that way. On the contrary, I readily took on any job that would make their lives easier, and much to my father's surprise, I did it with joy. The smell of beef being chopped in the grinder was such a sweet aroma I find it comforting to this day. Oftentimes, my dad would hand over a paper-thin slice of raw beef and say to me, "Go ahead, taste it." Just like that, no salt, nothing. Actually, for health reasons, my dad is very much an anti-salt person, so he never even allowed salt at the dinner table at home. It was a constant tussle between him and my mom. Sometimes she would get up and quietly rush over to the kitchen counter to sneak a little salt in her hand, but by the time she reached the table, Dad would grab her hand and say, "Nope!"

"Just a little bit," she'd counter.

"Alright, a little bit," he'd concede, and she'd sprinkle a dusting over her plate and throw the rest into the sink. It was always a big joke at the table.

"Mom," I'd say, "you cooked the food, why don't you just sprinkle some on your plate before you sit down?" It was almost like she wanted to get caught. I think it was their little game.

When he offered me that thin slice of what was basically carpaccio at the butcher table, I was in heaven.

I'm not sure if my dad realized that his plan to discourage me from being a butcher was not working, but I certainly wasn't going to bring it to his attention. I didn't care about the early morning wake-up calls and the constant freezing temperatures, because once I started working, I instantly felt wide awake and warmed up. I learned that it was a big mistake to sit down in that deep cold because it would quickly begin to settle in your bones, so I was always on the move, which made time fly by. Sometimes it felt like the clock was ticking way too fast. I didn't want the shifts to end. That sense of urgency that came with supplying meat to those eagerly awaiting their orders was like nothing else.

When all the orders were finally out the door, my dad didn't sit down and smoke a cigar; he kept going. He'd start fixing things around the building, and I'd shadow him like an apprentice. Pipes would freeze in the winter, so he taught me how to hold a flame to sweat a pipe, fluxing it with a pipe cutter, rejoining it, and then pressure testing it. Once I got shocked while splicing two wires together, and he immediately nicknamed me the Splicer. Jokes aside, I am very grateful he passed on all of these skills because being a jack-of-all-trades is one of the secrets to owning and running your own business smoothly and it allows you to keep the budget in check in the process. These lessons weren't just for me. Many of my dad's employees have shared stories throughout the years of him passing on not just butcher skills but also electrical and plumbing skills. One guy said he managed to do all the plumbing in his house because of my dad. He runs a tight ship, but I never saw him scream at any of his employees; there was always room for joking around no matter how stressful the day, and he always took the time to help where help was needed. All of this created a wonderful sense of camaraderie that was amazing to see and experience for me as a young kid—another valuable business and life lesson. That's also likely one of the many reasons so many people love and respect him. He's a good man.

By the time I hit thirteen, my dad let me get on the delivery trucks as a helper. I was super excited because that meant getting to drive around the city and learning yet another aspect of the business. It provided me with the best field education yet. Why was being on the delivery truck so important? Because it gave me access to the restaurants, the kitchens, and the chefs. Because of New York City driving and parking nightmares and to avoid tickets, oftentimes the drivers would remain in the vehicle and send me out to make the final delivery. Aside from finding this incredibly exciting, it was also the first time I began to interact with chefs. I'd pay close attention to what they liked, listening and relaying feedback to my dad, who'd never really developed these relationships. In turn, I was able to learn even more about our industry. That's also where I picked up fantastic cooking tips; it was just beautiful to see how a raw steak transformed into a sizzling masterpiece before my eyes. As the years rolled by, I may have stopped hopping on the delivery trucks, but I never stopped visiting those kitchens. I have lasting relationships with many chefs—

they show me what they want, and I come back with options and suggestions. My phone line is constantly open to them and I'm always at the ready to make their lives easier when they find themselves in a cooking bind. And all of that I learned from simply being a delivery truck helper. Never underestimate the potential of what might seem like a menial task.

I also developed a thicker skin. Chefs don't mess around, they don't have time to sugarcoat their words—they will tell it like it is. I had to learn how to grin and bear it, because it's part of the job: the client always comes first. Once, a restaurant owner who was also one of my dad's longtime customers, catapulted into such an explosive rant that after I stepped out of that meeting on that late afternoon, instead of going home, I went to the movies. It's the only time I walked into a movie theater by myself, but I was in desperate need of some time alone to clear my head. Everything was resolved the next day, but the truth is this is not uncommon. Some chefs can reach their boiling point in a matter of seconds, and even if they go ballistic on us, we have to simply breathe deep and take it. At the end of the day, it all comes down to getting them what they want, within their price point, and on time.

The West Village of the 1980s and '90s was a far stretch from the fashionable, clean, touristy area of today. Being open at three in the morning meant that anyone could walk through our door; friends, boisterous and brazen chefs fueled by a night of excessive drinking, amicable yet slightly edgy mobsters, and thieves. Yes, we got robbed at gunpoint a couple of times, but nothing major happened. All in all, we had a pretty safe run, and sometimes it was quite the spectacle.

There was a garbage company with trucks coming in and out just south of our building, and a Yellow Freight shipping company on one of the corners with a flat parking lot next door, which housed its tractor trailers. The backs of those trailers faced our entrance and provided us with unexpected nightly shows. The neighborhood cross-dressers would pick up men during their late-night walks and use these trailers as their hotel rooms. Good thing for them, cell phone cameras didn't exist back then. Then there were the newbies who didn't know that if they pushed those trailer doors down too hard, they would automatically lock up from the outside, leaving the couple trapped within. Those were the mornings when we'd hear banging and screams of "Help!" "Let me out!" On those occasions, the cops were called in, and even though what was going on was pretty obvious, they drew their weapons before carefully sliding those doors up, only to reveal the expected. Other times, the tractor trailer drivers who came from out west (some even donning cowboy hats) became inadvertently involved in these rendezvous.

The satisfaction, the stress, the tedious tasks, the phenomenal three-in-the-morning entertainment, you name it, I loved it. But my father wasn't having it. I knew I couldn't go against his will, so I went off to college. I figured if I wasn't able to follow my dream of becoming a full-fledged butcher, then I would follow his dream for me and become what he considered to be the pinnacle of success: a doctor. That's how I ended up as a premed freshman in 1989.

It was the day before the start of my first college Christmas break that life came crashing to a standstill. Prior to heading home for the holidays, excited to rejoin my family after completing my first semester at Albright College, I quickly called my dad at the shop, and one of his employees answered.

"Hey, Pino, let me speak to my dad."

"What do you mean? Your dad's not here."

"Well, where is he?"

"He's at the funeral home taking care of everything for your grandfather."

Shock coursed through my body. I went numb. I just couldn't believe my ears, but it was true. My parents had decided to keep my grandfather's passing from me because they knew it would be paralyzing. They were just trying to protect me from receiving such devastating news when I was alone, miles away from any family comfort. Plus, they knew I was coming home just in time to make the first night of the funeral. My dad was furious when he found out I knew. And ultimately, I realized they were right. I don't even know how I managed to get on that bus home—it was one of the most gut-wrenching moments of my life. It was an enormous loss for my entire family, even more so for my dad. To this day, he still goes to the cemetery every Sunday to visit his father's grave—he's that type of sentimental guy, with an enormous heart.

As the months went by, I noticed my dad was beginning to seem depleted, while my dream of joining the business only deepened. Shortly after my grandfather's passing, I decided to join the army as a combat medic. We come from a long line of patriots, so, aside from proudly wanting to serve my country, I also thought this would help my future medical career. A career that never came to be because those science classes were killing me; I just couldn't keep up. By the end of my sophomore year, after finding out I had failed Organic Chemistry, I was out. I quickly redirected and chose finance as my new major, killing my dad's dream of having a doctor in the family, but I didn't quit the army. My basic training began in the summer of 1990 when I was nineteen years old, and I went on to serve in the army reserve for six years, then added two extra years, and was honorably discharged in 1999.

My father was not happy with this decision—it made him nervous. Looking back, I understand. He had just lost his father, so the possibility of losing one of his sons in war was particularly distressing. My basic training coincided with the start of the first Gulf War in August 1990, so there was a chance I might be shipped off, though it never happened. Another close call came in 1995, when the United States got involved at the end of the Bosnian War, but Operation Deliberate Force was over almost as soon as it started, and the war ended a few months after that. Much to my dad's relief, I wasn't sent out to that one either, but that didn't stop him from worrying.

Being in the army reserve meant that I spent every weekend training at the Lancaster, Pennsylvania, base. Little did I know that, yet again, the invaluable education I was receiving would eventually influence the way I ran my business. It actually went both ways. Some of what I learned from watching my dad lead at LaFrieda Meats influenced how I led in the military, and some of what I learned in the army made its way into how I lead our business today. As a combat medic, for example, we were in charge of prepping troops before they shipped out to Bosnia. Every soldier needed to get a shot beforehand, and that clearly fell under our responsibility at the base, but we didn't have enough needles for all the soldiers. So we used an air tank to replace the needles, which meant air would shoot the medicine directly into the muscle, a very painful process. There were two or three of us set up to deliver these shots to the twenty thousand soldiers getting ready for duty. When I looked up from my station at the sea of people before us, I quickly realized that if we didn't change up our strategy, there was no way we were going to make it through everyone. I stopped all of the soldiers on my line and shouted out, "Take your BDU jacket off, roll up your sleeve, hold it there, and when I say 'Step forward,' you step forward. I'm gonna hit you and you're going to step away to allow the next soldier to step forward." And then I simply started yelling out, "Step forward! Step forward! Step forward!" By the end of the shift, I had gotten through three-quarters of the twenty thousand men and women just by making some adjustments to the line. That effectiveness came from my years working at LaFrieda and trying to make our lines efficient there, in a place that was slowly becoming too small for our growing business. It was training I didn't even realize I had.

Additionally, the army's sleep deprivation training still helps me get through my sixteen-hour work shifts, and the Primary Leadership and Development course I took to become a sergeant taught me how to deliver messages. I remember one drill in that course that consisted of receiving a paper with an object drawn on it and having to describe to the twenty-five people in the class what was on the paper without showing them the drawing, which was usually some crazy irregular shape to make the task more challenging. The grade was based on how many people came close to interpreting the image based on our description. The lesson really came down to learning how to translate a message as efficiently and effectively as possible no matter who was in your audience, something that serves me well today both with relaying specs to my butchers on the production floor and interpreting my clients' needs.

By the time I was in my twenties, I was a sergeant with more responsibilities than the average guy, leading a team of soldiers, some of whom were older than me. That's when I had the chance to test a theory that came straight from our shop's production line and the way we do business. To avoid having employees drag out the night because they need overtime, everyone automatically receives ten hours of overtime. That means that if they have to work the extra hours, they're covered, but if they're done early, they can go home and still get their OT pay. In the army, it wasn't about the money; the value was in our time. So, when my team completed the day's task, I let everyone go home. The incentive here was getting done early enough to have extra time to spend at home. Sometimes, a commanding officer would come around and ask me where my troops were, and I'd reply, "All are present and/or accounted for, sir." It was the truth—they might have already been home enjoying themselves, but I accounted for them because they had gotten the work done. It was stressful, but I thrived in those scenarios. This type of motivation has served as one of my keys to getting through whatever task is thrown my way. In the army, this once pushed me to lead my team to build a field hospital in around six hours when it would have normally taken eighteen. The hustle now for my employees is to get done early, go home, and not have to worry about dragging out useless hours at work just to get some overtime, which means they get their money and my clients get their meat on time.

The army involved doing something meaningful that had nothing to do with compensation. Serving my country was fulfillment enough. It was a stark contrast to my first job out of college. After graduating with my finance degree, I passed my Series Seven exam, became a certified stockbroker, and started wearing a suit and working on Wall Street. I knew my dad expected me to put my college degree to good use, so I put my dream of taking over the family business on hold and continued down the road my dad thought was the right one for me, but I hated every minute of it. The company that hired me, which was pretty large, was up to some sketchy business, but I didn't know any better at first. I was selling intangibles to strangers over the phone, paying for leads, reaching out to people who had already been called dozens of times in one day, trying to drum up new business. I did this all from my desk, lacking the personal interaction I so cherished at LaFrieda Meats.

I started to feel something was off. Why were we pushed to dump worthless stocks on innocent clients? It just didn't add up and made me extremely uncomfortable. About ten months after joining this company, I quit. I couldn't take it any longer. We weren't making our clients any substantial money, we were really just making money for the company, and that rubbed me the wrong way. I felt like I was doing something wrong. I collected my personal belongings from my desk, headed to the elevator, and left that life for good.

Even though I knew it went against my dad's wishes of me doing something bigger and better with my life, my dream was to do that something bigger and better with our own company. Why waste my time making money for others when I could do it for my own family? This is what I had wanted to do since I was ten years old, this is what my dad had unwittingly trained me to do on all those days when my time off from school was used to help out at LaFrieda Meats. I always felt that if our family business died, it would almost be like my family had died. I was determined to continue our legacy. My love for the business only grew stronger with time, and I had suddenly reached a now-or-never moment that urged me to put those years of invaluable training, my college degree, and my passion to the test. The only obstacle I still had to face: my dad. The time had come to go for it.

# 7

# PACK, DELIVER, GROW, GROW, GROW

ith my finance degree under my belt, my position as an army reserve, and out of a job, I approached my dad and expressed how much I wanted to join the family business and expand it. At the time, he was only selling to forty-four restaurants and working five days a week. I recognized the inefficiency, the lack of marketing, the lack of ingenuity. All the components were present—it just needed a fresh take, someone to reorganize all the moving parts for a better outcome. But my dad turned me down, yet again. "There's no money in this business," he'd say. "You'll be rubbing pennies together. I didn't send you to prep school and college and spend all that money on your education for you to be a butcher." He made a solid point. I have two brothers and a sister, and our dad spent all of his hard-earned money on our education for us to have bigger and better opportunities in our lives. I also know he was trying to protect me from the vicious competition and an industry that he thought didn't have much room for the growth that could generate a prosperous future. He just didn't see the potential that I could visualize, and

I knew I needed to get in there to be able to actually show him what we could do.

I decided to reach out to my aunt for help. Aunt Lisa was like the mob boss in our family, and I knew that if anyone had the power to get through to my dad, it was her. She lobbied for me, laid out all the benefits of having me on board, including my vision for our growth, and as his trusted adviser, she managed to convince him to give me a shot. By the end of that day in 1994, I had the green light. I was finally where I wanted to be.

Maybe my dad thought this would be a temporary arrangement until I landed another finance job. What he didn't realize was that I had spent most of my college years quietly observing and thinking up ways to grow the business. I was thrilled at finally having the chance to show my dad what we could do. Now it was my mom, my dad, and me, together with two employees—that was it. Meanwhile, the Fourteenth Street Market had begun to lose its grip on the market as industrialization advanced, an opportunity in the making that we had to be ready to pounce on to begin our own expansion.

In those years, competition had become so fierce that companies would do things like give each other flat tires to stall deliveries or climb onto the roof and cut phone lines—especially on Thursday afternoons, so that come Friday morning, you'd walk into your shop and have absolutely no messages with orders on the answering machine. It was ruthless; they did anything to bring you down. There was nothing worse than getting to work before dawn, ready to process the night's restaurant orders, only to find an empty answering machine. What do you do from three in the morning until the chefs wake up and get to work at noon? Nothing. We needed to make contact to find out what we had to cut for them. It was nail-biting. We'd try to keep busy with other tasks and prepped our delivery trucks so that we could fill them up as soon as we discovered what the day's orders were, but having lost so many hours meant that our drivers would then have to head out into peak New York City traffic, and under those circumstances, there's only so much you can do.

Potentially losing customers because we couldn't fulfill their requests, especially when we prided ourselves on always coming through for them, was a total nightmare. It was such a nerve-racking experience that one of my first orders of business when I officially joined my dad was to loop three answering machines together and set them up in the office so that if one failed or the tape was full, the next one would kick in. That wouldn't prevent possible cut phone lines, but at least we could ensure that when the lines were open, we had everything in place to receive those messages. Those requests are our opening bell, signaling us to start the day.

Happily ditching my Wall Street suit for my beloved jeans, flannel, and fleeces, I jumped right in to work, holding my knife as if it were an extension of my arm, ready to cut and portion the product of the day. Then I swung into high gear and helped pack the orders and load them onto the truck, which I would later drive to make the final delivery round, all along making mental notes on how to improve our efficiency. However, to really start growing the business, what I needed to focus on was reigniting my relationships with the chefs and restaurant owners and finding more clients. Up until then, my mom had taken on the role of visiting restaurants around the city, leaving our business card, and then taking the calls at the office. It had been great, but I knew we could do better. After my delivery rounds were over and I was back at

LaFrieda, I'd rush up to the bathroom and change out of my jeans and shirt and into my suit—I still had to wear a suit, but I didn't mind it as much because it was now in service of my company. With one outfit change, I went from butcher to businessman and hit the road to get new accounts. It took time and energy and lots of patience, but I loved interacting with people face-to-face and communicating my passion for what we had to offer.

We've always liked working with smaller restaurants not only because of the personal connection that we can create but also because restaurants come and go, especially in New York City, so if you have only one enormous account and that place happens to shut down, it could mean the end of your business. Diversification has always been key for us, even to this day. I think it's part of the reason we are still standing as a family-run company after decades in this industry, with clients ranging from small local restaurants to chains like Shake Shack. That's why I was so excited when I landed my first big account. I walked into Becco on West Forty-Sixth Street and there was Joe Bastianich, the owner himself, arguing with his executive chef over the rising price of veal shank. Unlike most other restaurants, Becco serves the classic Italian dish osso buco year-round, so I knew this was my big shot. I respectfully jumped into the conversation, introduced myself, and mentioned that I was selling veal shank for a dollar a pound less than what he was getting it for. He was surprised and suspicious at first, thinking it was too good to be true, but he gave me a chance. I rushed back to tell my dad the good news.

"Who the hell is that?" he replied, always a little reticent when it came to something or someone new.

"Dad," I said with a muffled chuckle, "it's a big account. He serves osso buco all year, so he needs a lot of veal," and I went on to explain the details, even though I honestly was still learning how to pronounce Joe's last name properly.

The next day, we prepped his order, packed it, put it on the delivery truck, and it was all ready to go, when the driver called in sick. That meant I had to take his route. When I arrived at Becco, I pulled my LaFrieda hat almost over my eyes and tucked my head way down low, praying Joe or the chef didn't catch a glimpse of me. What would they think if the same guy who was in a suit selling them on our company was now in jeans and a T-shirt delivering the goods? It wasn't the professional message I wanted to relay, but we had no options. That's how small we were back then. Thankfully, neither of them was in that early in the morning. Years later I came clean and told Joe I used to make the deliveries too. "Get out of here," he replied.

As I slowly started bringing in new clients, I knew it was time to upgrade our ordering system. Back then each restaurant had its own book. So, every morning my mom would listen to the messages on the answering machine and write each of the orders down in the corresponding restaurant books, by hand, using a carbon sheet between two pieces of paper, to create an automatic copy for our records. If by chance that carbon sheet went missing, she would write the orders out twice, once in the books, and once on a sheet that we could take with us to the production area. I knew right then that this was no way of growing a business. We needed to fix this, we needed to digitize our system. However, there had been so many stories going around for years about systems crashing and companies losing all of their data and going out of business because of their missing accounts receivables that my parents weren't willing to take that risk. Their philosophy was more along the lines of "If it ain't broke, don't fix it," so changing their minds was a bit of a struggle. I can't tell you how painstaking it was to work hard all morning only for a restaurant to call at two in the afternoon and tell us we had delivered the wrong meat or made a mistake with the specs. I needed to fix that, and they eventually got over their fear and conceded. It was perfect timing because upgrading our system and purchasing accounting software was the logical next step to sustain our business and open our door to the expansion and growth I envisioned.

Back then, the software didn't cover everything I needed. The meat industry is extremely specific. Most accounting software has an inventory aspect to it, but it's of no use to us because we have primal cuts coming in and then a slew of smaller specified cuts going out, and we just can't inventory that with existing software. As technology has advanced in the last years, I have been able to work with coders to edit the software to meet our needs. So, for example, nowadays, when a restaurant places its order, if it has ten items and three need to be butchered, the software will take those three items and create individual butcher labels for each one, including the necessary specs, date, and USDA stamp information. It's been a steep learning curve, and there's still room for improvement, but we've come a long way since the mid-1990s. The ability to log specifications for each restaurant not only saved us time but also improved our accuracy and efficiency.

The growers send cattle to the harvesting facility, which in turn processes the animals and produces the primal cuts—including the chuck, brisket, rib, loin, and round—vacuum packs them, and sends whatever cuts we've ordered to our premises. We then receive the product, unpack the beef, and run it all through peroxyacetic acid, a type of diluted vinegar that inhibits the growth of pathogens and bacterial contamination on contact. Ten years or so before, we were still using the N60 sampling program, which consisted of procuring sixty small individual samples from all around the big bins of meat to test them for pathogens. Those bins were not allowed to enter the building until the test results were returned. Then we tried lactic acid, but to our dismay it turned our beef gray, especially the chopped beef, and that's how we eventually landed on peroxyacetic acid. It keeps our beef safe, doesn't change its color, and leaves no aftertaste. Like my dad says, that machine is the least expensive one in the building because a single recall with our name on it could cost us our entire business.

While the fresh meat goes through this lifesaving machine, restaurant and retail specs for each order are being printed and distributed to our team of butchers. The primals are then sent to their corresponding tables to get cut per each spec. That's where the magic happens. If a chef asks me for fourteen-ounce, single rib bone rib steaks, I have to go downstairs and teach my crew how to split the bone to produce the client's request. Usually my production manager and right-hand man, Manny Campos, and I take a few nights to train the team of butchers until we feel confident enough to let them do it on their own. That happens with every special or out-of-the-ordinary request. Our team must be trained to read and correctly interpret the spec labels, so they'll know exactly what is required. It's a time-consuming process, but it's a testament to our attention to detail.

Once the cuts are finished, if the restaurant doesn't require their orders to be individually packaged, they are paper-wrapped and the butchers add an adhesive label with the product's information. If it's a retail order or a restaurant that does require individual packaging, then it goes to the packing area, where each cut is individually packed and labeled. When it comes to our burgers, once the boxes are filled with patties, each box is set on a conveyor belt that goes through a metal detector. If there happens to be any metal in there, a loud beep will go off and the belt will automatically stop. This is just to ensure nothing fell into the burgers when the beef was being chopped and the patties formed. I'd say 98 percent of the time, when it does go off, it's usually a staple on the box, so we proceed to remove the beef from the box and run it through again to make sure it's all clear. In addition, I've now installed a density machine, so if there's bone matter that's dense, it'll pick it up, since small fragments of bones are potential problems. It's easier for them to get through the system, and it's a complaint that we get every so often that I would like to eliminate entirely.

When all of this is done, everything is sent to the warehouse, where it gets weighed, confirmed, and shipped. I like for us to have one last check at the warehouse because if there's a packaging mistake, that's often where we catch it. Back when I started out on Leroy Street, every piece of meat that left our facility went through my hands. I'd personally walk to the scale, weigh the items, pack them, double-check everything, and then take them out to the trucks. To think we only had one scale back then and now we have fourteen.

Every step of the journey within our facility is overseen by the watchful eye of a USDA inspector. The beef we cut, chop, or grind at the start of each shift is organic, so that it's not contaminated by any other beef. The rest of the night we usually deal with the all-natural product, and we close the shift with Standard and Utility. Our USDA inspector checks to make sure that we are effectively transferring the quality grades and any other certifiable specifications and pertinent information from the labeled primal cuts we have received from the harvesting facility to the final subprimal cuts we package for delivery. The Food Safety and Inspection Service oversees label development and compliance and enforces beef labeling laws. Falsely labeling products is unlawful and can be punished with warnings, fines of up to $10,000, and up to three years imprisonment. That's right—if you add something that the USDA hasn't verified or approved, you run the risk of going to jail.

The USDA inspector's role has been continually evolving since I started in the business. Good working relationships are important to this day, of course. But the previous system made it difficult for the meat company to survive if you weren't able to make friends with your inspector. Sometimes an inspector would walk in all high and mighty and write a facility up for no reason, just so it looked good on paper for his boss. If an inspector didn't like how his (and I say "his" because there were only male inspectors back then, that I recall) day was going, then he could close the operation, on a whim. We saw a few places completely shut down like this—that was our biggest fear back then. Think about it: a stranger basically had the power to walk into our business and access any part of the building they wanted—an intrusion and abuse of power—and write us up if they thought something wasn't up to code, no questions asked, no arguments allowed. That grew extra hairy when you had to deal with an inspector who abused alcohol or drugs, which skewed their objectivity and led them to write places up based on other motivations rather than the strict observance of the rules.

The situation had everyone on edge. We had no choice, like everyone else, but to get over this frustration. Sometimes we forgot the looming consequences, like the one time at Leroy Street when my dad was speaking to my grandfather and an inspector tried to interrupt them to get something signed. They were absorbed in their conversation, and I watched as the guy tugged at my dad's butcher coat a few times to get his attention, and my dad turned around and went off on him about his lack of respect for interrupting him when he was speaking to his father. It's one of the only times I've ever heard my dad scream, and to a USDA inspector no less, who is basically a government agent with a badge and all! Things could've gone south quickly, but that guy played it cool and walked over to a corner like a reprimanded puppy, patiently waiting for my dad to finish his conversation. There was so much going on back then that I'm guessing my dad may have known something about this inspector that made this guy react in such a docile manner.

It was strange because my dad has no violence in him at all. We say that he's a lover not a fighter. But he was on edge in the mid-1990s, and only when I started working at LaFrieda again did I notice the growing tension between him and my mom. Up until then, my parents had always been close; I had never seen any issues growing up, so the strain I was perceiving was unexpected and new to me. At first, I chalked it up to work stress. My aunt had recently retired, my mom was the bookkeeper, and my father and I were cutting, packing, and delivering, all while I was trying to grow the business. We were spread thin—every person in that building was absolutely essential to stay afloat. Meanwhile, the USDA inspector stationed at our facility, who we were on good terms with, suddenly started writing us up and handing out an unordinary amount of NRs (process deficiency records). That meant that we constantly had to interrupt our workflow to take care of whatever issue he cited as wrong, then show that we had resolved it, and have him sign off on the repair before we could continue working in that area.

To put it into perspective, last year I only received a few NRs in the entire year. Back then, we faced a period of about four months straight where we were getting written up often. Most of the NRs happened to be about issues on the second floor, far away from my mom's first-floor office. We later found out that while we were upstairs fixing whatever menial disrepair was on the list that day, the inspector was downstairs flirting with my mother. Tensions flared between him and my dad, and when all was said and done, the inspector left, never to return, and my father and mother began their divorce proceedings, which were likely already a long time coming.

For a moment there, I honestly thought we might not make it. The following few months, we had to survive without a bookkeeper. This guy had abused his power, but we couldn't sue the government, and suing him wouldn't get us anywhere anyway. So we had to accept the turmoil and forge ahead. It took me a few months to find an accountant who was the right fit, so I did my best to fill that role until we finally found Maria, an amazing, highly competent person who really helped get our business back on track. She worked with us for about thirteen years, until she passed away. Thank goodness that even though we had already doubled our roster from forty-four to about eighty clients, we were still relatively small. Nevertheless, handling this personal and professional upheaval was overwhelming. To top it off, on Thursday nights I met with the Menu Club in Queens—an organization that brought together a representative from each food industry sector to talk business and new opportunities over dinner—and every Friday I had to leave from Leroy Street and travel three hours to Lancaster, Pennsylvania, for my weekend army reserve training. I'd get back to New York late Sunday night, sleep a few hours, and be back at work before dawn on Monday. It was incredibly taxing. However, despite having a tough time when it came to some of the following inspectors, who were tight with the previous guy, we managed to survive. And eventually, as they got to know us, the relationships smoothed out to the point where I'm still friends with a lot of the inspectors who came through our Leroy Street doors.

I've seen USDA meat inspection transform since then. Like anything else, those bygone mishaps were corrected over time. It's a much more professional setting now. What happened to us wouldn't happen today. They wouldn't take that risk. Nowadays, we have the utmost respect for our inspectors, and that fear of impending doom has subsided. At the end of the day, when the role is fulfilled properly, it makes complete sense. I'm grateful for the USDA and its team of inspectors. They keep our facilities up to code and our industry and end consumers safe.

On a personal level, the next few years were difficult, but eventually the wounds healed, and my mom returned to work with us once a week. However, seeing what my parents went through taught me a hard-earned lesson I was determined to put to use in my own life. Working with family is already hard enough, but working with your partner all day can be taxing on any relationship. I've always kept my wife and two kids separate from my work life. I have enough on my plate dealing with the stress of working with my dad and cousin Mark. I love them both to death, so the last thing I want to do is argue with either of them because of differences on the job. It happens, of course, and that's one of the most difficult parts of running a family business. I hate going home after having argued with my dad—the feeling of unrest lingers all day until we are able to resolve the issue. Thankfully, as we grew and got busier, I had to start work earlier, my dad remained on his three-in-the-morning shift, and Mark works the day shift, so now we overlap with each other a few hours a day, but we're no longer together nonstop. It's all about finding the right balance.

Meanwhile, my plan to keep expanding our business did not falter. The biggest challenge was getting my dad on board with change. My dad has always been my role model—I will forever look up to him with deep respect—but when our elders are set in their ways, well, making upgrades or implementing any form of change can be an uphill battle. We'd butt heads left and right as I set out to prove my dad wrong on a number of things just to get him to evolve with the times. Usually, once he was able to see that the upgrade actually worked or made our life easier, then it was smooth sailing. So, I quickly learned that talk wasn't going to get me very far. I just had to take the plunge and hope for the best. And yes, that meant making mistakes too. Like the time I tried our luck at home deliveries. Once again, I suffered the consequences of being ahead of our time.

It was the mid- to late-1990s, right smack in the middle of the dot-com boom, and it dawned on me that if we set up a website exclusively for home delivery, we could start shipping our product across the country. Excited by this prospect, I reached out to a friend—a West Point graduate who was a tank commander in the first Gulf War and a computer genius—and shared my idea. He got to work building the site. My dad's hate for retail butchery had not waned with the years, so you can imagine how opposed he was to this idea, but I pressed to prove my point. I bought $10,000 worth of insulated boxes and stashed them on the second floor, but they were corrugated and porous, which meant they quickly accumulated dirt. First learning curve, but I survived.

Once the site was up, the orders started pouring in. At one point we had to shut it down because it was growing beyond our means. And then came the real nightmare. We had just shipped a decent amount of product for home delivery on our biggest weekend of orders yet, right when UPS, our carrier, decided to go on strike. That meant that our product sat on a loading bay for two weeks until it was finally shipped. And then the calls started streaming in. I had an older couple from the Midwest on the line explaining that they had just received their delivery and noticed that there was brown juice dripping from the bottom of the box, and they wondered if it was still okay to eat. I was flabbergasted. Credit card reversals, refunds, endless customer service calls, you name it, we did it. It was a nightmare and a big loss for our business, all due to bad timing. The right systems and protocols weren't in place yet to handle this type of delivery, and the strike obviously didn't help. If I had waited a few more years, it would have worked beautifully. And don't get me started on having to deal with my dad's I-told-you-so looks. Eventually, we took another stab at it and put the site back up just to get the phone volume down, and it has been running strong ever since, now managed by my brother Chris. Every night, UPS comes by to pick up a few pallets for home delivery, and during the height of the COVID-19 pandemic, it was a true lifesaver on all fronts.

One of the things that allowed me to even dare venture into home delivery when I did was the existence of vacuum packaging, which simply means manually or automatically removing air from a package and sealing it shut. In the beef world, this means meat can be sealed in an airtight package and kept fresh and safe for up to ninety days. The technology can be traced back to the 1940s, used mainly during World War II to provide soldiers with food that could last for days on the battlefield. But the machine for home and industrial use was invented by German Karl Busch and brought to market in 1963, revolutionizing countless industries. However, not all were willing to follow suit. As this technology grew in popularity, butchers and chefs were mainly against it when it came to meat. Oxygen on collagen helps break down the muscle fiber and makes meat more tender. So when meat is vacuum-packed, the air is taken out of the package, meaning there's no oxygen to break down the collagen, so it prevents any type of aging process that would tenderize the beef. On the flip side, it preserves the meat for a much longer time, which opened the doors for harvesting facilities to ship their product longer distances without worrying that any of it would go bad.

Previously, meat hung on hooks was shipped in refrigerated trains and trucks, and was only good for a couple of weeks. Compare that to ninety days and you will understand the industry's excitement over this development. Eventually, the butchers and chefs hopped on board because they realized that now the harvesting facility also had the ability to section out the muscle groups for each industry sector. So, for example, we no longer had to buy entire foresaddles and commit to selling all its other sections just to get our rib eyes. That changed everything for us, the Meat Market, and beyond. It also meant that the roast beef clients, for example, no longer had to take in hind quarters that were useless to them and only created more waste. That waste is now sent directly from the harvesting facilities to rendering companies. This satisfies other markets while saving the butcher or meat purveyor money because they no longer have to spend time trimming excess fat and paying for it to be removed from the premises. Furthermore, we are now able to purchase what we need knowing it is good for ninety days, and we've gone from my grandfather and dad simply wrapping meat in a bag and only being able to make local deliveries to vacuum-packing our own product and shipping it anywhere across the country. This is the beef industry's efficiency in action once again, something that will never cease to amaze me.

Our first vacuum-packing machine became a hot commodity when Angelo Ponte, the mobster who for years controlled New York City's garbage and sanitation, was sentenced to federal prison. Turns out he was on a special diet that required he get food that was packaged at a USDA facility. Ponte's, his restaurant on West Street, was one of our clients, so his son arranged to bring his father's special food over to Leroy Street and my brother and I vacuum packed and labeled it for him with our small in-house machine. I'm talking stuffed pork loins with broccoli rabe, pork crown roasts, lamb roasts . . . I wish I was on that special diet! One day, they brought us the meals and included some Italian lard bread, and when we vacuumed the air out, the big loaf went down like a cracker. We looked at each other in a panic: "Holy crap, we have to fix this before he comes back!" But there was no turning back, so when Angelo's son came for the meals, we showed him what happened, and he took it lightheartedly: "Alright, you guys keep that, don't worry about it," and he grabbed the rest of the food and left. We took it out of the package and went about our business only to find that after about thirty minutes, the air permeated the loaf and brought it back to its original state. My brother and I love Italian lard bread, and we were famished, so that was the best surprise of the day. Of course, we later explained that it did work and it was only a matter of waiting for the air to do its magic. To think that we now have a $200,000 vacuum-pack machine in our facility that can seal up to thirty-three steaks a minute.

As my dream of expanding LaFrieda began to bear fruit, others in the industry took notice, and one day, in the late 1990s, a guy who knew my dad reached out to offer me a job. He was a successful owner in the process of merging his three meat distributorships and, after seeing our increase in business, he asked if I would want to come work for him. I took the meeting in his second-floor office on West Street out of curiosity and following a philosophy that I still stand by: always take the meeting. Even if you aren't going to accept the offer, those meetings can be a treasure trove of information. The offer I got that afternoon: $150,000 a year. For a twenty-seven-year-old, that's a lot of money, probably four times what I was making at the time. I carefully listened to his pitch and then thanked him and politely declined. What he didn't know was that I had already landed my dream job and no amount of money was going to change my mind.

Then came offers from other companies who wanted to buy our growing business. We received one such pitch when I was twenty-nine. I remember it clearly because they were offering $29 million, and all I could think was, *A million dollars for each year of my life . . .* but that didn't even come close to derailing me from what I was doing as the owner—selling was out of the question. However, when you own a business, that's not a door one should close completely because if things go south, it could be the only shot you have to survive. Never close the door, and always take the meeting. These types of offers give insight into the value of your company from an outside perspective. The investment company that was behind this first offer was Bear Stearns, one of the banks that tanked during the housing and financial crisis only a few years later. I shudder just thinking about that what-if.

I've also had a number of companies approach me to see if we'd be interested in buying them out, which has taught me how to evaluate businesses and gauge if it's a solid opportunity or not. But I'm getting way ahead of myself. Let's go back to the 1990s. With Pat LaFrieda Meat Purveyors in full expansion mode, I really needed my weekends to run this growing business, so, feeling satisfied with having given nine years of my time to my country and with no wars in sight—or so I believed—in 1999, I was honorably discharged from the army reserve. The time had come to lead our business in full force, and I was more than ready. But I quickly realized I couldn't be everywhere at once. Spreading ourselves so thin was no longer an option. So I approached my dad with the idea of hiring a salesman.

"Dad, we need a salesman. I can't do all of it. I wish I could, but I can't."

There were only twenty-four hours in a day, and I devoted four to sleeping.

"If you hire a salesman," he replied, "then they'll expect a commission."

"Yes, Dad. They're not volunteers."

That's when I thought about my cousin Mark Pastore. He is an amazing people person, everybody loves him, and aside from considering him one of my best friends, he knew the business inside and out and was an incredible PR guy. It was a no-brainer, and he jumped right in. However, a short while after Mark quit his other meat establishment job and came to work for us, we got a visit from some guys connected to that other company, one that was heavily controlled by the mob, as most of the meat businesses were back in the day. My dad had one steadfast rule: we should never be associated with the mob in any way. Sure, we crossed paths with them, and as Italian Americans we became friendly with some of their members, but we always kept them at a safe distance. So when these guys came calling, I knew we had to figure out a way to get out of this unharmed. I refused to give them Mark, which was their initial request, so eventually we settled on a trade: Mark could stay with us if we agreed to buy three pallets of poultry from them every week, at fair market value, which amounts to about six thousand pounds. And I said yes.

It was a win-win because I was already looking into other poultry suppliers. We had ventured out of the Meat Market when my father negotiated with Cooking Good for them to bring us a half trailer of their product every Tuesday or Wednesday, which amounted to about eleven pallets, or twenty-two thousand pounds of poultry. However, when they were bought by Perdue, the minimum amount they allowed was an entire trailer—way beyond what we could move. We had to decline, and that's where we stood when these guys presented this trade for Mark. Aside from doing anything to get these guys off my cousin's back and also actually needing the poultry, I knew this particular company was in trouble and it wouldn't be long before they all went away—it would be a temporary arrangement at most. Three months later the FBI came looking for the owners for participating in organized crime and the company was forced to close. Mark has been with us ever since, and now he's our business partner—we wouldn't have come this far without him. He's been a key factor in our growth and success, the best purchase we've ever made.

Another important person at LaFrieda, whom I introduced earlier as my right-hand man, is Manny Campos. He used to work for the same company Mark was at before joining us, has been in the industry for twenty-five-plus years, and has been with us for around fifteen of those years. Manny came to work for us right around the time we started to have real, tangible growth spurts. "I had seen the LaFrieda trucks and knew they bought from the Meat Market, but when I started out in the industry, they weren't anyone's real competition," explains Manny. He witnessed first-hand how cramped we were at Leroy Street in the early 2000s. We were working elbow to elbow in a facility that was starting to feel like a horizontal Jenga game, where we moved one pallet to get another one out just to move the next one to the side. Manny came in as one of four butchers, and now there are close to twenty butchers on the floor every night. He doesn't even have to cut meat anymore because he's in charge of managing them. He also knows how far we've come with our delivery trucks.

We started out with a small truck and a beat-up old van on its last legs, then grew to three trucks by the time Manny joined us, but by then I already knew it wasn't enough. One night, I was meeting with one of my clients, August Ceradini, the then COO of World Yacht and chairman of the Culinary Institute of America, and I complained that my trucks were getting out late because we were growing faster than expected. He just looked at me and said, "So? Do something about it." What was holding me back? The constant clashes with my dad. Every time I suggested we needed another truck and another driver, he thought I was exaggerating. He is very conservative when it comes to spending, and that's actually a huge plus in the bigger scheme of things, but when it comes to growing a business, change and investments are par for the course. So I took August's advice and went for it anyway. I got another truck and another driver and dealt with my dad's complaints for a day or so until he realized life was easier and he piped down.

Now we have fifty-two trucks delivering meat to restaurants and they're all equipped with GPS and a tracking system so that when a customer checks in asking about their delivery, we can tell them exactly how far away that specific truck is from them. After all, we sell meat, but we also sell service, and our goal is to always come through for our people, no matter what.

**SEPTEMBER 11, 2001,** began like any other Tuesday morning. We had already finished cutting all the meat to fulfill the day's orders over at Leroy Street, which was only a few minutes away from the Twin Towers. My dad had just left in one of the trucks because he was training a driver that day, and I was taking a brief nap before hitting the streets to make the daily sales rounds. That's when the first plane hit. As the news broke, I rushed outside and witnessed the building in flames. Soon after, I watched as the second plane struck and I panicked—this was no accident. One of my dad's first deliveries was for Jubilee Market on Gold Street, only a few blocks away from the towers. I desperately tried to reach him on our Nextels—a type of walkie-talkie. Those few minutes of silence pushed me to the edge, only to be relieved by the sound of his voice on the other end. He was already on the Verrazzano Bridge en route to Staten Island. Amid the rush of emotions, for a brief moment, trepidation overcame the horror. Two days earlier, we had finished expanding our building to allow room for an additional freezer and refrigerator. I had just spent all the money we had to make this move, and now, as I watched the chaos on the streets unfold, I kept thinking, *I can't believe I just blew our family savings*. It was hard to picture Manhattan ever recovering from such a hard blow. But as I quickly learned amid the turmoil, we were an essential service. Our shop was closed on Wednesday, back open on Thursday, and as busy as ever. New York still needed to be fed, and we had the access to get it done. The same was true when Hurricane Sandy hit more than a decade later—all the bridges and tunnels were closed except for trucks delivering food to the city, so we turned our generators on and kept working through the mayhem. It's what we do.

Knowing that people need me to come through with a delivery and depend on me, well, that automatically kicks my adrenaline into high gear. I thrive on those moments—they make me feel useful and alive. Beats the hell out of selling intangible stocks to strangers over the phone. Here we're talking food that people eat for pleasure, to nourish themselves, to live. I'm all about that—I absolutely love coming through for my customers. Does that mean sacrificing other parts of my life? Sure, it does, but it's part of my DNA. Funnily enough, all the hard work and long hours and peaks of stress have actually helped keep me centered through life's ups and downs. I recently realized that starting at the age of thirteen or fourteen, I had likely been suffering from anxiety and depression, and the only constant factor that gave me the necessary steadiness to get through it all was focusing on my family business. I learned to use work to manage these emotions without even realizing it. I've grown out of it now, and maybe that's why I feel comfortable enough to openly talk about it.

I remember one of my first fears in life was a nuclear war with Russia. The existence of this possible threat was hammered into us as kids at school. I took it so seriously that every time I heard a loud explosion or saw a bright light, it sent my heart racing and made me shiver.

Another moment that deeply affected me was when I was mugged at thirteen. To celebrate my good grades, my grandfather had given me a new bicycle, and I was over the moon. What thirteen-year-old boy wouldn't cherish such a gift? As I was riding it down the street one day, I felt this redheaded kid speed past me on foot. I watched him run by, in a daze, wondering if I could ever run that fast, when suddenly he turned around and came charging back in my direction. Before I could even grasp what was happening, he had beaten the heck out of me and taken my bike. In shock and in pain, I hobbled home but latched on to hope because, for some odd reason, I believed that crime victims were reimbursed by the state or the country. When I walked through the door and told my mom what had happened, I asked her, "So, how long will it take for the government to send me a new bike?" She looked at me, puzzled, and said, "What are you talking about?" The days that followed, no matter where I went, I kept looking over my shoulder, fearing this kid would come out of nowhere and deck me again.

The real anxiety kicked in when I was about fifteen. But it never crossed my mind to hurt myself. I never got to that point. I just remember as a teenager, being out with friends and having a great time and all of a sudden being overcome by a crushing feeling of guilt for not doing something more productive, like studying, followed by an immediate urge to go home. So I wasn't fully able to enjoy certain experiences like other people.

Everyone deals with a certain level of anxiety from time to time, but in my case, when it got excessive, I channeled it into my work—that was the one place where I felt focused and in the game. Pushing myself to work longer and harder helped me blow off steam. It also played a role in our growth through the years—unlike many of my competitors, I was always available and accessible, because I was always at work, and that made a big difference for clients with last-minute requests.

Later on in life, when the anxiety finally fizzled away, it was like a buzzing light had finally been turned off—once the noise disappeared, its existence suddenly dawned on me: I had suffered through all of that without even knowing what was going on. Now, I sometimes observe my dad and wonder if he hasn't suffered from anxiety himself and never realized it, because he's always worried about something. I focused all my energy on our company, so much so that I think at times one of my dad's biggest worries was *me,* because I could border on obsessive. In the end, though, it really helped me cope. Work, more specifically going down to one of the tables and cutting meat all night, still fills me with a sense of peace like no other. It's my meditation. That's where I find my center, that's what clears my mind and allows me to get ready to face whatever is next in line. The butcher table, the knife, the meat—that's home.

8

# THE BIG LEAGUES

Burgers. I can still remember those giant four-ounce doubles I grew up eating as a kid in our Brooklyn backyard, grilled to perfection by my dad, topped with a simple slice of American cheese, and embraced by sesame-seed bread we bought from the corner bakery. As I bit into that perfect piece of American comfort food, I could never have imagined that this would ultimately be what catapulted us into becoming a household name in New York City and beyond. It all began with our chopped beef, the one my great-grandfather Anthony started selling in his shop in the 1930s, a signature blend of chuck, brisket, and short rib, the same one my grandfather and dad continued selling in the city. The secret was to take it seriously. This isn't about grabbing meat leftovers and grinding them into oblivion until they become a nondescript patty. Like my grandfather wisely said, "You can't hide your sins in chopped beef," a rule to live by at LaFrieda.

The family blend called for the same amount of specific beef cuts to create a uniform flavor that our clients could count on. That's how we first came to be known for our chopped beef in the New York restaurant circle. The beauty back then and to this day is that we use whole cuts of different muscles, many of which are usually used for steaks, and no fat trimmings—that pink slime is completely out of the question. This allows us to know the exact flavor profile we are dealing with, which makes our end product consistent. There's nothing worse than falling in love with what you believe to be the best burger you've ever tried only to go back and not enjoy the same thrilling experience. That is unacceptable. Furthermore, following how my grandfather did it, our meat is chopped at very cold temperatures, which keeps it coarse and prevents it from getting overworked. And you already know we pride ourselves on sourcing that wonderful, all-natural, domestically raised Black Angus beef, keeping with our high-quality standards.

All of these ingredients placed us in the perfect spot when the burger craze began to take hold of the city. The first client to approach me was New York restaurateur Henry Meer. For his soon-to-be opened downtown steakhouse, City Hall, he wanted something different, a sweeter, fattier burger than what he already had come to love from our original blend. I immediately knew I had to up the amount of brisket to get him the buttery flavor he was after. Once we hit the perfect proportions, we were all set. It was 1998, a pivotal moment for our business, our first custom blend and the first time a restaurant put our name on the menu, right under the word burger: Pat LaFrieda's Custom Blend of Short Rib and Brisket. Other custom blend requests soon followed.

The first question we ask before we even start to make customization suggestions is: What's your favorite steak? Each chef's palate varies, so hitting the nail on the head in terms of taste really comes down to what they like, and then, what they're looking to accomplish with their particular burger. Many times, we're asked for something with a rib eye flavor profile. We also offer grinds with different levels of coarseness, which changes the taste and plays another key role in creating a unique blend for each restaurant.

Then came 2004, a milestone year for us. I had already been working with restaurateur Danny Meyer, providing customized chopped beef for his Union Square Cafe burgers, when he approached me with his latest idea: opening a small burger stand, or shack, in Madison Square Park. He needed something that wasn't too dry yet not too fatty to avoid a kitchen fire. Additionally, it had to be their own special blend, something that would make them stand out among casual burger joints. I went back to my grandfather's original hamburger recipe, and after some back-and-forth with Danny, I took his feedback and tweaked the blend until he was completely satisfied. I also told him this had the added benefit of being an all-natural blend with no antibiotics or growth hormones, something he could easily advertise to his customers. And he hit a home run. Soon after they opened, I took my dad to Madison Square Park to see what all the fuss was about, and the scene was jaw-dropping. We had never seen such a long line of people willing to wait for an hour or more just to get their hands on a burger. It was staggering, and it changed our lives.

Having such a sudden and unexpected smash hit, about six months into their opening, Danny came to me again and asked if we could form their patties at our facility. Dad gave me a stone-cold NO from the start. He was a staunch believer that the best burgers had to be formed by hand. But this was a huge client for us, so no wasn't an option.

"Dad, they want us to form four-ounce pucks. What if we were to buy a forming machine?" I said, hoping he would see reason. Pucks are basically patties that are extra tall so that when they hit the flattop grill, the cook can press them down and get that mouthwatering outside layer of caramelization.

"What the hell, we're butchers, we don't sell hockey pucks!" replied my dad, annoyed.

I let it go because I still had to figure out how to form 1,200 four-ounce burgers a day without a machine. I resorted to biscuit cutters first. A team of five people filled these cutters with Shake Shack's blend to form the pucks and then placed them in boxes for delivery. I had this small operation going for weeks, if not months, to my dad's constant and loud opposition: "What are we gonna do next," he'd grumble as he walked by, "cook it for them?" He remained resolute, but I knew our little operation needed a major upgrade in order to continue supplying these pucks on time. Despite my dad's stubborn refusal, claiming that it would cost too much money and we didn't even have room for it in our Leroy Street shop, I went ahead and ordered a customized burger-forming machine. It was a risk, a leap of faith, but I had to take it because I knew I couldn't let this opportunity pass us by.

I started measuring the space where I thought it could go and ordered a $150,000 custom-built machine that I designed myself just to form those specific burgers for Shake Shack. Danny and I had signed no type of formal agreement, so if he happened to pull out from the deal, I would have been screwed. But as the saying goes: No risk, no reward. The machine was delivered during the day right when I was out doing my rounds to see clients. When I returned, I walked over to the meat inspector, who was smoking a cigarette outside, and asked, "What's going on?"

"Some asshole ordered a burger machine that doesn't fit through the front door," he replied.

My heart dropped. I had measured everything down to a tee,

including the height within the building, but the front door never even crossed my mind. The hopper, which is where the meat is loaded, was simply too tall. I stood there, filled with dismay, staring at this large machine planted outside our building and trying to calculate what I could do to make it fit, and then I saw my dad.

"Did you buy this?" he asked, slowly making his way toward me.

"Yes."

"Then you deal with it."

He was fuming, but I had no time to waste. I quickly dialed the manufacturer in Germany, explained the situation, and had them talk me through how to carefully remove the hopper from the rest of the machine without disrupting its inner gears and then reassemble it all inside. The rest is history. And just like with everything else, as soon as my dad saw it in action, he loved it. Now we have seven, working twelve hours a day, six days a week.

After Shake Shack's runaway success, our phones kept ringing off the hook with restaurants asking for the Shake Shack blend. That was not an option. Their blend was top secret. But I told each caller, "Let me know what you've got in mind and I'll come up with a special blend for you." We created several new blends for different restaurants, each one taking up to a few months to perfect, but our crowning glory came in 2008.

I had started playing around with the idea of adding some of my intensely flavorful dry-aged beef to a new burger blend. Something that could deliver the elite indulgence that comes with that expensive piece of meat at a price that would be slightly more accessible to the general public. At around the same time, chef Riad Nasr from Minetta Tavern reached out to express his desire to add the ultimate burger to their menu, something with a decadent flavor profile unlike any other burger anyone's tasted before. It also had to be an exclusive blend made just for them. And that's how the Black Label Burger was born, named by Mark, who got his inspiration while driving by a Johnnie Walker billboard.

People thought I was crazy. The stock market had crashed, we were in the thick of what would later be called the Great Recession, and there I was grinding up my prized dry-aged beef to create an expensive burger experience. What those who doubted its success didn't realize was that the customers who had been spending fifty dollars on a dry-aged steak were all too happy to have the chance to still indulge in that intense flavor at half the price, and those who couldn't afford the steak were now able to treat themselves to the burger. The following year, Minetta Tavern sold around thirteen thousand Black Label Burgers, and it eventually became one of the most iconic burgers in America.

The burger trend only grew from there. Now we have more than one hundred exclusive blends that we process every week, each unique to the corresponding chef's needs and palate, each blended individually and loosely packed to ensure extra flavor and juiciness, with end consumer price points ranging from six to thirty dollars. It's something that fills people with joy and transports everyone to backyard grilling with family and friends. At least that's where my mind goes every time I bite into a burger with that perfect bun to beef ratio and that explosion of flavor that awakens all my senses, bringing true comfort.

**THE BURGER PHENOMENON,** the burgeoning client list, the extra machinery, all of those roads led to one solid and obvious conclusion: we needed a bigger building. My Jenga-like moves in our Manhattan location could only get us so far. It would keep me up at night, wondering how I would logistically make it through the next work shift. The experience taught me how to use every last square inch of space, which continues to be of great service to me today. But there comes a time when things can get so cramped they become inefficient, and that's no way to run a booming business.

It's very hard to grow in this industry from scratch like we did. We had to strike the balance between being conservative and making the necessary upgrades to continue expanding, but we did so slowly and steadily, following one of my dad's biggest principles: be conservative with your profits and constantly reinvest in your company. To exemplify this point, Dad likes to tell the story about a competitor who had the edge on him back in the day. One Christmas, the competitor's three business partners got one another BMWs while my dad and Aunt Lisa invested their hard-earned cash in buying the properties along Leroy Street. They're out of business now and we're still thriving. That really cemented my family's philosophy. Stay modest at home, put your money back in your business, play it safe, and make sure there's always an extra cushion saved up to pay the bills and stay in the running if the economy takes a turn for the worse. When cash flow becomes a problem, it's very difficult to dig yourself out of those holes. These rules we live by have really helped us survive through the decades and become more competitive. And, sure, I made big purchases, as I've mentioned earlier, but I always made sure we made that money first so I could pay in advance.

Our success and our need for more space added up to my dad finally embracing the reality that it was time to move. I've talked endlessly about the Fourteenth Street Meat Market because that's where we started out and conducted our business, but Manhattan had another meat market in Harlem too. It closed down before I joined my dad in the business, but the buildings were still standing, some unchanged, which meant that they had the meat company rails and the infrastructure we required. Logically, that's where we first turned as we contemplated our big move. Our first prospect with these characteristics was located on 116th Street. It fit most of our criteria, so we were seriously considering making the purchase. Yet, when we did some due diligence, we found that this building was located only inches away from Columbia University, which meant that it could easily fall under eminent domain. That was a deal breaker for us—we weren't about to put ourselves at risk of having our building taken away from us at any point, leaving us a few million dollars in the dirt. Far from a sound investment, I know we dodged a bullet there.

Next, we turned to Google Maps, which had only recently launched. We opened the site on our computer, typed in New York City, and stared at the map on the screen, wondering where else we could contemplate moving. The idea was to go forward, not back, so we discarded Brooklyn, which was where my great-grandfather initiated our family into the business. Queens, well, to get over those bridges, that traffic was way too tough and not something we wanted to put our drivers through, let alone have them run the risk of not making our deliveries on time, so that borough was out. Traffic from the Bronx into Manhattan was intense too, and that's when we zoomed in on North Bergen, New Jersey, located on the other side of the Hudson River, right across from Midtown Manhattan. We didn't know the area, but when we realized the amount of space we could acquire, we were astounded. Add to that what I had been trying to tell my dad for a few years—that Manhattan just wasn't conducive to processing meat—and we decided to look into it further.

Leaving Manhattan, hell, leaving New York, that was far from an easy decision to make. Logistically, it made sense. Emotionally, it was tougher to process. New York was everything to us. And Leroy Street was our home. The thought of losing our New York identity was painful, and leaving the city felt like a loss at first. Yet when we sealed the deal and broke ground on the new 35,000-square-foot property, I knew it was the right thing to do.

The new facility was designed from the ground up by my dad, custom-made to efficiently fit our every need. We can now process meat that feeds more than three hundred thousand people a day, make more than seventy-five thousand burgers a day, and have two dry-aging rooms that house more than five thousand cuts of meat, the equivalent of eighty thousand steaks. Did we lose our New York identity? Probably a little. But it sure didn't affect our New York business. We are now only a tunnel ride away from the city, within reach for our deliveries and our clients. I honestly thought one of the drawbacks of moving across the river would be that chefs would give up on their impromptu midnight visits because we were no longer in the vicinity. But I was wrong. We have more chefs stopping by now than we ever did in New York. They even ask for tours.

One night, around four years ago, I was in production when I got called into the office by one of my guys over the loudspeaker. I wasn't expecting any visitors, and chefs usually call me straight on my cell phone. Turns out it was a man named Aaron Mizrahi, who had just graduated from culinary school in upstate New York and was heading back to Mexico. All he wanted was five minutes of my time so I could sign his copy of my cookbook, *Meat*. Even though it was a busy night, I said of course, and went back to my spot in production to continue cutting meat. Around forty-five minutes later, they called me on the loudspeaker; he had already arrived. The meet and greet was supposed to last a few minutes . . . and we ended up speaking for four hours in my office. Fast-forward to 2019, and that one impromptu conversation led us to doing $16 million of business in Mexico.

To get to that moment, to build that kind of demand for our product in Mexico, took four years and relentless hard work. Aaron's dream was to bring higher-quality beef to Mexico. When we first spoke, I didn't even think it was possible to get our meat there, price-wise, because I didn't think there was a demand for that upper echelon of beef. The problem was that no one had ever entertained the idea, tapped into that demand, or supplied it to discover that there was indeed a market there, ready and eager to finally get some decent meat. A lot of beef that goes to Mexico is frozen, not fresh, and most of it is on the lower end of the grading spectrum because beef is really price-sensitive. However, Aaron connected LaFrieda Meat Purveyors to this one restaurant group, Sonora Grill, that is home to thirty-two restaurants in the country and was eager to add our name next to its beef dishes on the menu. The demand was stronger than I expected, but one question remained: How do I send millions of dollars' worth of beef every month to Mexico and make sure I get paid? I couldn't take such a huge risk without some sort of guarantee that the money would show up in our account. After meeting four or five prospective financial partners, and none of them panning out, I wasn't taking the possibility of this deal too seriously, until I met Geraldo, a financial partner who just so happened to also be a logistical genius. He was the missing piece.

When the deal was finalized, I flew to Mexico City, where Aaron helped take it to the next level with his media connections.

He set up a five-day media tour, we did around thirty-five interviews, television appearances, and radio shows, and ended the trip with an event scheduled to announce that Sonora Grill would be carrying Pat LaFrieda beef. Before flying over, I was asked if I wanted a bodyguard. Puzzled, I said no. Then my insurance agent called me. It just so happened my company insurance was about to expire.

"I have a few more questions to ask you, Pat," said my agent. "Will you be traveling abroad?"

"Well, I'm going to Mexico in a couple of weeks for business," I replied.

He paused, then said, "Oh, you're going to need kidnapping insurance then."

"No, I don't want that. Why?"

"It's a big problem in Mexico."

"Okay, if you introduce me to the guy who has the suitcase of money who's going to come save me, sure, I'll take it, otherwise I don't want it. And I'm going to tell everyone I don't have it because I don't want to get kidnapped for the insurance money." And that was the end of that. No kidnapping insurance, no bodyguard. No need to advertise what I thought would be a relatively low-key business trip.

What I thought would be a small gathering to celebrate this partnership turned out to be a full-fledged event with people lining up for a book signing and to take photos with me. I couldn't wait to tell my dad about the experience because he had had his doubts about the whole venture too. At the end of that evening, as I departed the venue with the owner of Sonora Grill, he asked me if I wanted a bodyguard to escort me to my hotel. All I could think was, *Why is everyone talking about bodyguards? I don't get it.*

But I simply said, "No, and I don't have insurance, just so everyone knows. No one is coming to save me. I already instructed my entire team not to accept any phone calls from any kidnappers." He chuckled. Meanwhile, I kept thinking, *Oh my god, I'm going to get kidnapped, this is it.* But I continued to play it cool and nothing happened.

When I traveled there a second time, to Monterrey, Guadalajara, and the Chihuahua region, I knew I needed a bodyguard. Aside from traveling in what is known as dangerous territory, I had something else on my mind. To get this business, someone had to lose business. Those companies were based in this region, and that's where I was heading. Additionally, I was becoming a recognized figure within that circle, which I found odd, but taking all of this into consideration, this time merited hiring a bodyguard. It was the sensible thing to do. The truth is we ruffled a lot of people's feathers with our deal both in Mexico and the United States. American beef companies didn't want high-end beef going to Mexico. I had to break relations with some of the ones who didn't see it our way or tried to prevent us from doing this. I still don't get why this was an issue to them, but what's important is that all the beef that gets to this group is our beef, our brand, and not the processing facility's beef. After all was said and done, I came back from both trips thinking Mexico was fabulous and absolutely beautiful. And I felt safe.

Following my two visits, a group from Sonora Grill came on a steak tour of New York City, hitting all the top steak restaurants in town and ending their five days with a meeting at my office. They didn't speak English very well, and I don't speak Spanish, so Aaron served as our translator most of the time. He basically told me they were disappointed in the restaurants they had visited because the culture of fine dining in Mexico has a nightlife component to it, which was lacking in the city. I explained that New York food critics would never write kindly here about a steakhouse with energy and music that gives it a slight clubby feeling. Nevertheless, when they noticed this void, they began considering opening a few locations in the United States and asked for my advice regarding what cities might work.

Now, Aaron and Geraldo are hell-bent on helping LaFrieda eventually expand down to Central and South America. First we need to finish ironing out the kinks of our Mexico operation. It's a work in progress, especially when it comes to controlling the beef quality, but that's part of the growing pains of a relatively new deal. Once we get the logistics and processing right, once it's running smoothly, then we can explore more options. We don't want to spread ourselves thin—I've learned that lesson the hard way. Like I tell everyone I work with: If everyone at the table does what they promise, we have a long, prosperous business relationship ahead of us. Nothing comes easy, but hard work pays off.

What's Pat LaFrieda Meat Purveyors' secret business sauce? We think things through, plan, learn from mistakes, and never lose focus on our priorities: meat and our clients. We are the last line of defense before the customer gets their beef, so we have to remain vigilant and keep improving our performance and efficiency. It's the way we've always run our business, throughout the generations, since day one, with quality, transparency, dedication, and loyalty as the cornerstones to our success.

# PART THREE

# TO YOUR TABLE

## 9

# WHAT'S THE BEEF WITH EATING MEAT?

First and foremost, let me make one thing clear: If you choose to be a vegetarian, lacto-ovo vegetarian, vegan, pescatarian, if you want to just eat blueberries for the rest of your life, that's your choice, and I completely respect it. I am no one to tell you what to eat or not eat. Everyone has the right to choose where they source their food and what they put in their bodies. So, why the constant animosity against meat eaters? What's behind the beef with eating meat? Let's start with an argument I've heard time and again: our bodies weren't made to eat meat. If you're nodding your head or looking quizzically at this page, rereading that line, wondering what to make of it, let me stop you right here. It's not like we woke up one day a few years back and started to devour this food group out of the blue. Humans have been eating meat on a regular basis for more than two million years.

Once we started eating meat, with time (we're talking hundreds of thousands of years) our digestive systems actually adapted to this food group and evolved. Since we no longer had to chew and process so many plants, our gut basically got smaller.

Additionally, all the energy that used to go to finding, eating, and digesting plants was redirected to the brain, an organ that's ravenously seeking nutrition to grow. By providing our bodies and our brains with a nutrient-dense source of food, not only did our guts become smaller but our brains became bigger. Then, around eight hundred thousand years ago, humans learned how to cook meat, which made it even more digestible and energy efficient.[1] It also meant we no longer needed large, sharp teeth to tear into a carcass. By processing meat, over the years, our teeth became smaller and our jaws less muscular, which may have led to changes in our skulls and necks that in turn made room for our growing brains and advanced speech organs.[2] So, in essence, it could be argued that eating meat is what made us the thinking and evolved humans we are and perhaps what gave us the intelligence to even have these discussions about what to eat or not in the first place.

Humans have evolved and we're constantly trying to keep up with how what we put into our bodies affects our well-being, which brings us to one of two main arguments used when promoting and defending plant-based diets: the health factor. Once again, let me make this clear: I am not against plants. They are an important part of a healthy and balanced way of eating—emphasis on balanced, which is what may be lacking nowadays.

Beef is a source of high-quality protein loaded with essential nutrients. When it is taken out of the equation in a diet, this nutrient-dense food must be replaced by several different foods to get the same amount of protein and to supplement its existing vitamins and minerals. The murkier waters come when plant-based proteins enter the picture, more specifically the plant-based burger craze, and the health and environmental claims that accompany it.

Are plant-based burgers really healthier than having an all-natural beef patty? I think it's been openly proven and accepted that they're not. Even a quick Google search comparing the burgers from the two latest trendy companies, Beyond Meat and Impossible Foods, with a beef patty will reveal a whole lot of extra sodium in the plant-based burgers (around 370 milligrams versus the 70 milligrams in beef). Then there's the ingredient list. These plant-based patties list around twenty ingredients versus the one ingredient in beef burgers: chopped beef. Furthermore, they are considered a highly processed food because they're made in a lab. Beyond Meat uses pea protein to mimic beef protein and claims to contain non-GMO ingredients, while Impossible Foods uses soy protein and genetically engineered ingredients. Neither of them is organic and both use methylcellulose and food starch, which can also be found in ice cream, sauces, canned soup, and so on, to bind their product. And that's about all the consumer will get to find out because both companies consider their recipes top secret.

This opens up another can of worms. Under the FDA's GRAS (generally recognized as safe) program, "substances that are GRAS under conditions of their intended use are not food additives and do not require premarket approval by FDA."[3] That means the USDA doesn't regulate vegetarian and vegan labels. Legally they can write anything they want on their packaging. In that sector, it really comes down to hoping that you are purchasing from a trusted brand that is not out to bend the truth in their favor. No federal oversight means no accountability. Since the FDA's approval is not required for these ingredients, it's up to the food manufacturers to claim, or "self-affirm," that the ingredients are safe. How do you know if what they claim you're getting is in fact what you're eating? It's hard being inspected, sure, but at least we have an entity holding us all accountable; it creates a fair playing field when it comes to quality and claims. And don't even get me started on "natural flavors"—anything goes under that category.

If we're focusing strictly on health, and you were given the choice between a patty formed from a long list of ingredients, with high amounts of sodium, created in a lab, versus one that has only one ingredient, more protein, less sodium, is nutrient dense, and has all-natural and organic options available to you, which one would you choose? When you have a meat industry that's highly regulated by the USDA compared to a plant-based product that isn't—well, I'd rather go with the one that has all of these checks and balances in place to keep the food we're consuming safe. What's more, aren't we supposed to stay away from ultra-processed foods and focus on whole ingredients? Then why turn to processed foods to replace a natural unprocessed option? Something just doesn't add up when it comes to the health argument.

This brings me to another point, a pet peeve related to processed foods and more specifically to processed meats. Not all meat is created equal. We've already talked about the different grades and certifications, but now I'm referring to fresh all-natural nutrient-dense beef versus processed meat. These are two different products, yet they are often lumped together as one in countless articles on how red meat affects our health. This is simply unacceptable. Processed meats are considered carcinogens, they require preservatives—usually nitrates, which are linked to cancer—and have up to four times more sodium than unprocessed meat. How can article after article put them in the same category as fresh meat, as if they were the same thing? They should be studied and reported on independently because they do not have the same effects on our bodies. We're inundated with so much conflicting information that it's hard to figure it all out on our own, and even harder when we realize that our go-to sources often don't know what they're talking about either.

**IT WAS 2003** and my aunt Lisa had just been rolled into surgery to take care of an ulcer. Thinking it was just a standard procedure, my father and I were taken aback when we saw her doctor walk into the waiting room and head straight toward us. As we stood up to meet him, we didn't know what to think. He cut straight to it: "Lisa doesn't have an ulcer, she has stomach cancer." Stage four. The metastasis was too widespread to even attempt to remove or treat the tumors. It looked like someone had opened her up and thrown rice in her cavity. Her lymph nodes were also infected. It was too late, irreversible. They gave her six months.

In the hopes that she could extend her life a little longer, she went on chemo and made it to month nine, but passed away in agonizing pain. Her death was a huge blow to my family and me. She was everything to us. Aunt Lisa had always been there for my dad and for me. If it hadn't been for her, my dad might not have given me the go-ahead to join the business. Seeing her undergo such pain and then losing her was beyond dreadful.

After she passed away, I became acutely aware of my own health. I was overweight at the time, definitely overeating, and suffered a distinct discomfort high up in my digestive tract after each meal. Something wasn't quite right, and I knew it, but up until then I had chalked it up to going overboard with my food intake. Plus, I had already been diagnosed with acid reflux, so I assumed it was related to that. However, after watching my aunt suffer and disappear before our eyes, I decided to go see her specialist for an endoscopy.

"Pat," he said, noticing the worry on my face, "this is not genetically linked. There's no reason to give you an endoscopy."

"It might not be genetically linked, but I have a problem," I replied.

We went back and forth for a while since he was convinced I was just being paranoid, and it got a little heated until he finally conceded and booked my endoscopy. I returned a few days later. Prepped and ready to get this over with, I waited for my anesthesia to work its magic. Did you know that 0.1 percent of people suffer from something called anesthesia awareness, meaning that even though they can't move or speak, they are actually awake, aware of their surroundings, and can even feel pain while technically under general anesthesia? I didn't, until that day when I fell into that 0.1 percentage. There I lay, unable to move a muscle, feeling that the doctor was pushing a baseball bat down my throat. And I could hear everything. I was alert, awake, enduring the pain, and silently begging for it to be over ASAP. Then my doctor, who had seemed calm up until then, suddenly blurted out, "Fuck, fuck, fuck." That's when I knew something was wrong. After they rolled me to the recovery room and the anesthesia wore off, the doctor sat by my bedside, and I simply said, "Doc, what's with these two nodules that you found?"

He looked at me in shock. "Who told you this?"

"I was awake through the whole process."

"Stop right there, don't say another word," he replied, and immediately stood up and left the room.

He came back shortly with two hospital lawyers, who were

holding a chart with those happy-to-sad faces with numbers below each illustration, ten being the saddest of them all. And they showed it to me and asked me where my pain had been during the procedure.

"Off your chart."

But I didn't care about their charts or concerns. I reiterated that I wasn't planning on suing them and then asked them to please leave. It was such a drawn-out process. Imagine having to go through this legalese while still in the dark about what's lurking in your body. I desperately needed to know what the doctor had seen and what he had found. I wanted him to get down to the facts.

"You have two nodules," he said, once the lawyers were out of the room. "I took biopsies of them both and did an endoscopic ultrasound." He wanted to see if the nodules had grown past the stomach wall, a sign that would indicate I was in deeper trouble than expected.

My endoscopy was on a Tuesday and the results were due Friday. There was no way I was going into the weekend without getting those results. That waiting period is pure torture—anyone who's gone through this knows what I'm talking about.

I finally got my doctor on the phone, and he said, "The nodules have grown into the stomach wall, but we've caught them in time." He explained that I had intestinal metaplasia (meta = change, plasia = cells), so the cells were changing into a premalignancy, exactly the same as my aunt's cancer, which unfortunately wasn't caught in time. The clincher: both her cancer and the threat I had growing in my own stomach were likely caused by the consumption of processed meats: hot dogs, bacon, prosciutto, cold cuts, all of that. Nitrates, nitrates, nitrates.

"Doc, nice knowing you," I replied after he broke it all down.

"No, hold on, you need surgery!" he said.

"Go fuck yourself!" I screamed and hung up. I was fuming and furious at that guy. To think I had to argue with him to get an endoscopy in the first place only to find out I had these precancerous tumors, which were the same type that had killed my aunt! If my aunt hadn't died, if I hadn't insisted on getting that test, if I had left it unchecked for longer, I might not be here to tell the story. No way in hell was I going to let that doctor treat me too.

After making some calls, I headed over to New York-Presbyterian and met with two surgeons. Unfortunately, this wasn't an easy get-in-get-out type of operation because one of the offending nodules was located close to the lower esophageal sphincter, the stomach valve. During these types of surgeries, normally the doctor removes the nodule and an inch and a half around it, what they call its margins, in case there are precancerous or cancerous cells still lying in wait in the vicinity. However, in my case, if they did that, it would mean removing my stomach valve. One of the surgeons reassured me, "I can do this without touching the valve." He said he could take three-fourths to one inch tops around the tumor close to the valve, covering most of the margins. When I asked what would happen if he just took the valve out too, he said he would have to place a type of rubber band there and my quality of life would diminish. "Whoa," I said, as that option sank in, "then let's go with the three-fourths of an inch." It was a risk I was willing to take, and thankfully it worked.

The surgeon removed both nodules and I remained in the clear for the next couple of years. Then, at one of my six-month endoscopy checkups in 2007, they found a new nodule, so I had to undergo another surgery. Thankfully that last one was far easier to remove than the first two, and I've been okay since then. I went from half-year checkups to every year, and now it's every two years. Stomach cancer is irreversible, but by catching it at the metaplasia stage I have been able to beat the odds.

Suffice to say, after that odyssey, I no longer include processed foods in my everyday diet. Do I have a piece of prosciutto every so often? Yeah, maybe. Do I eat a hot dog every now and again? Seldom. Listen, I grew up eating Boar's Head bologna—that's basically the same as hot dogs, the same preservatives, everything, except bologna comes in slices and hot dogs are tube-shaped. My aunt Lisa grew up eating hot dogs. Back then, who knows how much nitrate was in our food since nothing was really tracked at the time.

That's why I know so much about these devils in disguise, and that's why it bothers me when people bunch processed meat with unprocessed beef. They are two different products with decidedly different health consequences.

**THE BOTTOM LINE IS THAT** I have nothing against meatless companies. I'm not out to knock them down at all and I have nothing against people eating alternative burgers or any other type of processed foods. I'm just seeking the truth, and trying to make sure we, as consumers, are well informed rather than hypnotized by claims that are spread by the media with numbers skewed to fit their stories.

It's important to note that Beyond Meat and Impossible Foods are not primarily focused on advertising that their products are healthier than beef; their angle is that they're sustainable and better for the environment. Beyond Meat claims that, when comparing their patty with a beef patty, their patties generate 90 percent less greenhouse gas emissions and require 46 percent less energy to produce. This is based on a study conducted by University of Michigan's Center for Sustainable Systems. I haven't been able to find any independent studies backing this up, so I choose to take the colossal difference with a grain of salt until more research emerges. The fact is that they're processing these burgers at plants and using industrial oils, such as coconut oil. If this oil isn't domestically sourced, are they factoring that into their final carbon footprint? Only time and independent studies will tell.

Why am I so skeptical? How would you feel if the technology developed to make a similar end product to your own just so happened to be backed by Big Tech? Coincidence? I bring this up because I want the honest answers before assumptions are made about what's good and bad for the environment and our health. I think there are a lot of nuances that are eagerly ignored. If these companies have a much smaller carbon footprint, why did their product cost around thirteen dollars a pound when they launched? It's not like they were making it for a dollar and selling it for thirteen. There's a reason it cost so much, but the process is not transparent. Furthermore, there's a cost issue for the average consumer. A family that is struggling to make ends meet can't afford to pay thirteen dollars a pound for these meat alternatives. When activists push for a meatless world, that means taking away the option of getting a natural unprocessed source of nutrition that costs around four dollars a pound, three times less than what they proposed as a solution. Not everyone can afford what the plant-based activists are suggesting everyone should eat. Most people can't even afford basic organic food staples.

You want to talk about sustainability? A meatless world is not sustainable. It doesn't take into account the more than seven hundred million poor people in the world,[4] or those who are not officially poor but are teetering on the low-income line. It also doesn't consider that many families across the globe rely on basic agricultural activities to survive. According to the UN Food and Agriculture Organization, livestock alone contributes to the livelihood of 1.7 billion poor people and 70 percent of those employed in the sector are women. Livestock provides these families with an income, a form of transportation, and a resource to help them produce crops, all of which plays a big role in reducing poverty and battling food insecurity and malnutrition.[5]

Could a meatless world population actually survive? The answer is a resounding no. Not at the rate we're growing. Our current 7.8 billion world population is scheduled to hit approximately 9.7 billion by 2050 and could reach 11 billion by 2100.[6] So aside from providing essential work and cash flow, we need meat to help feed the world. My company alone feeds hundreds of thousands of people a week, and I have the demographic and price breaks needed to hit all sectors of beef consumption, so that no one is left out.

With all this being said, I do support plant-based protein. As you can see from the numbers above, we will need every available source of food we can get to feed our growing population, and meat alone will not be enough, just like meat alternatives alone wouldn't cut it. It's not about eliminating sources of food, it's about adding to what we currently have. We need both, not because one is healthier or more environmentally friendly than the other—we ultimately need both to survive in the future. Additionally, since we will at some point not have enough commodity beef to cover the world population's needs, alternative protein will also keep our meat prices from skyrocketing.

That's why, when Beyond Meat first entered the food scene, I was one of the first companies to endorse them. Almost four years ago, the CEO invited me to a special dinner where they unveiled their product. When they described the technology, I loved the innovation behind it and started to carry and distribute their product. To help them grow the business, I suggested they build an East Coast facility, change their entire marketing strategy, and have restaurants serving their product. And now we carry and distribute Impossible Burgers—a fact that always comes as a shocker when interviewers throw out their favorite "gotcha" question asking what I think about these alternative meats.

Theoretically, if any improved product hits the market, and my customers are asking for it, that's a plus for me. It's something else for me to carry and distribute. One doesn't have to cancel out the other. If you have a new technology to get protein out of wood, I would support you. While I can taste the difference between a meatless burger the real thing, I have had plant-based protein as a chopped meat substitute mixed into other dishes created by top chefs, and I can honestly say I couldn't tell the difference. It does a great job as a meat substitute when it isn't featured as prominently. I see its potential; however, I don't see it as competition. There will always be a demand for meat. Personally, I don't want meat to go back to being just for special occasions or for the affluent, a luxury item the average American can no longer consume. It needs to continue to be available and affordable to all.

I'm not here to preach to anyone what to eat or not to eat. That's a decision that should be made with the right information at hand. And if you choose to not eat beef for health, environmental, or ethical reasons, I respect that. I'm not going to try to convince you otherwise. I simply expect the same reciprocity toward meat eaters. We can and should be able to coexist. At the end of the day, I want everyone to be happy in life and with what they eat.

## 10

# BRINGING BEEF BACK TO ITS GLORY

The first beef I ever cooked was filet mignon tips and tails, when I was around fourteen years old. While most of my friends ate Steak-umms, I was playing around with filet mignon. To us at the shop, these were, in essence, simply trimmings. When cutting filet mignon, the tip and the tail ends become too narrow to sell to restaurants as a proper filet, but they're just as tender as the prized cut, and easy to cook on a skillet, which meant it was something I could be trusted with indoors. In the summers, when I was home from work and my parents were away for the day or the weekend, my siblings and I were left to fend for ourselves. Well, not entirely. My mom always stocked us up with plenty of food to cover our meals and then some. But with no one in the kitchen, it was a chance to experiment and learn how to cook for myself, which I thought was super cool. That's how one summer day, left to my own devices at home, I grabbed some of the filet mignon trimmings we had brought in from the shop, seared them in a skillet, and used them to make a steak sandwich. I'd seen my dad make this before with skirt steak, but never with the filet mignon trimmings. Little did I know then that this home concoction aimed to satisfy the incessant hunger of a growing teenage boy would eventually become the inspiration for Pat LaFrieda's Original Filet Mignon Steak Sandwich.

This toasted custom-made baguette stuffed with filet mignon seared to perfection, topped with caramelized onions, and finished with Monterey Jack cheese first made its way to the public back in 2012 at the Citi Field baseball park in Queens. When the Wilpon family, then owners of the New York Mets, approached Mark and me to ask us what we thought about contributing a steak sandwich to their menu, the sandwich I used to make at home immediately came to mind. They wanted something that would rival steak sandwiches sold at other ballparks. But rather than focusing on the classic Philly cheesesteak, I wanted to take it up a notch. This was also a great way to showcase our high-quality meat, bring awareness to our brand, and hopefully turn the baseball crowd into LaFrieda fans too. We'd already helped glorify the burger; now it was our very own filet mignon steak sandwich's turn. And it worked . . . it was a home run.

Our Citi Field stand was the first foray into cooking up some classics for customers. The next big call came in 2016. Stepping into the restaurant side of things for us means maintaining the delicate balance between offering some of our product for people to have a taste while not competing with any of our esteemed restaurant clients. There are no contracts that forbid us from doing this; it's just a question of what's right, of loyalty toward our longtime customers. When we were approached about adding a Pat LaFrieda restaurant to the Pennsy Food Hall on the corner of Seventh Avenue and Thirty-Third Street next to Madison Square Garden, once I confirmed we wouldn't be competing with any of our clients, I was in. This was our first retail restaurant, and rather than go the steakhouse route, we continued our sandwich theme—something our clients didn't really focus on, which meant it was safer territory. We featured our original filet mignon sandwich, and it was a hit, the most frequently purchased item on the menu. The exposure we got from that one location was invigorating. The goal was to not only provide the customers with a good bite to eat but to also supply them with an unforgettable meat experience. This in turn helped our name get out there, it drew curious customers to our website, and then to our clients' restaurants, bringing our brand exposure full circle and making it a win-win for all.

It worked so well that when Time Out Market offered us a spot at their new Dumbo location, which was scheduled to open in May 2019, we went for it. Once again, we decided to focus on building up our signature sandwiches, burgers, and what we called the World's Greatest Hot Dog. And once again it was a hit. So much so that in less than a year, we opened two other locations at the market. Turns out people still can't get enough meat.

It's no wonder that beefsteak parties became a thing in the late nineteenth century. They started out as a gathering of working-class men who would don white butcher's aprons and chef's caps, sit on crates or kegs in a dungeon-like setting such as a cellar or tavern, and tear into all-you-can-eat beefsteaks with their bare hands, washing them down with rivers of ale. No plates, no cutlery, just a napkin, an apron, the steak dipped in melted butter and served with a piece of bread, and mugs of beer. Maybe it started out as a break from all the primness and properness of everyday life. A place where these men could let their hair down, so to speak, forget about manners and impressions, escape life's doldrums, and simply enjoy a juicy steak in the company of like-minded guests. By the time the 1900s rolled around, the popularity of beefsteak parties had continued to grow and they had expanded into different sectors of society. High-ranking business leaders, politicians, and club members—including such figures as the Bloomingdales, Theodore Roosevelt, Mark Twain, and the New York Giants baseball team—all hosted or participated in these beefy gatherings, sitting at bare-boned rustic communal tables, delighted by the lack of refinement. It was a men's-only gathering until around the 1920s, when beefsteak dinners became the highlight at benefits, elite club events, and fundraisers, hosting society's crème de la crème. With hotels and restaurants replicating the original dungeons, the era of guests using their hands instead of forks and knives was mostly retired, as were the unlimited pours of beer during Prohibition, but beef remained the star of the show.

Over the decades, people have ventured beyond the sirloin steak into other cuts to fit each occasion, whether it be an important event, a big family gathering, or dinner at home with your loved ones. The following are some of the most popular cuts, including my favorites, to serve as inspiration for your next get-together.

Let's start with the classic porterhouse steak. The origin of its name is still disputed and will likely never truly be known. Some say it came from the Porter House Hotel in Georgia, others claim it's from Porter's Hotel tavern in North Cambridge, Massachusetts, whose owner's name was Zachariah B. Porter, and then there's Martin Morrison's legendary Porter House on Manhattan's Pearl Street, which many point to as the early nineteenth-century birth of this cut's name. Wherever the name came from, this is the father of all steaks, a favorite for many people, and one of my go-to cuts when I want to impress a restaurant chef because it has the strip on one side and the filet on the other.

Another one of my go-tos when wooing chefs is the classic rib eye steak. Packed with exquisite marbling, this tender cut is bursting with rich and juicy flavor. The boneless rib eye is often referred to as the Delmonico steak, after New York's first à la carte restaurant, Delmonico's, which opened in 1837 and quickly became known as New York's fine-dining pioneer, offering one of the first à la carte menus and separate wine lists in the country. After surviving closures and revivals, it continues to operate at its original location on the corner of Beaver and South William Streets.

Then there's the bone-in rib eye, my personal favorite, also preferred by many other butchers, mainly owing to the pleasure of stripping off the last bites of delicious sparerib meat straight from the bone.

If you want the meatiest ribs, go for plate short ribs—the beef is a little tough, so it's an ideal cut for braising or marinating, which helps it reach its tender potential.

Craving a rib eye but can't afford to splurge? Then go for its budget-friendly alternative: the chuck eye steak. It has the same tenderness and flavor as the rib eye and comes from the same muscle group—the front end of the shoulder—making it a juicy option for grilling.

The bone-in strip loin steak, aka the New York strip, is another chef favorite, and my preference too. Some people can't be bothered with digging around the bone for their meat and prefer the hassle-free boneless strip. But for me, cutting around the bone is part of the allure of the meat-eating experience.

An underrated beef cut that many say rivals the New York strip when it comes to flavor is the flat iron—I'm not sure I'd go that far, as there's nothing quite like a New York strip, but it's definitely a great budget-friendly option. What originally hindered its performance as a steak was a long sinew running down the middle. That's why for years, flat iron was mainly used as chopped meat. However, the flat iron is now cut into two filets and the sinew is removed. It has become a trendy and economical choice you will see on many menus across the country.

I can't fail to mention the world-renowned filet mignon, which in French means "dainty filet." This small boneless steak that comes from the tenderloin is by far the most tender cut of meat due to its delicate and tight muscle fibers, but it's missing the marbling that gives beef that unforgettable flavor. That's why many filet mignon dishes involve sauces that, combined with the tenderness of this cut, level up its palatability. However, my favorite way of eating tenderloin is raw. I slice it thin and revel in its buttery texture, reminiscing about those days as a kid when my dad used to feed me this thinly sliced raw cut while we worked over on Leroy Street.

Outside the production room, if you were to stop by my dad's house or my own on any given weekend, what you would most likely find on our grill is the outside skirt steak. It is my number one go-to beef cut. I've cooked so much of it in the week I'm writing this—at home, on Zoom calls, at butchering demos—that I think I may need a little break from it . . . but not for too long. If you think of Sammy's Roumanian Steakhouse, that was the steak to have. It's a classic that stood the test of time. On a personal note, outside skirt steak is a cut that my grandfather and dad have always loved, so there's an immediate sense of home connected to it for me. When I take that first bite, I feel like I'm in a safe place. But aside from that personal bond, I believe it's the one cut of beef that really stands out from the rest—it's easily identifiable even in a blindfold test. Although it may not be as tender as some of the aforementioned cuts, it's got that river mineral rock flavor to it, a little livery yet sweet, that sends me into a realm of pure joy.

And then there's that glorious bone-in rib steak that makes many people drool: the majestic tomahawk. The cut comes from the tomahawk rib chop, which has seven rib bones that must be evenly portioned out into seven steaks. Once cut, its main visual feature is the extra-long rib bone emerging from the beef like a mini skyscraper. It is revealed by frenching (aka trimming) the meat and fat surrounding the bone, transforming it into the lamb rib chop's extra-large brother. If you look at this cut of beef and compare it to the tomahawk singlehanded ax, you'll know where it got its name. Nothing like a tomahawk steak presented on the plate to impress your guests, and if you want to glorify the experience even further, make it dry-aged to hit them with that profound flavor that will ravish their senses.

This brings me to one of my favorite subjects: dry-aged beef. It all began in the late 1990s when Becco on West Forty-Sixth Street asked me for dry-aged porterhouse steaks. Knowing how quickly they went through those steaks, I said yes, recognizing the great opportunity before me. That was some of the first beef we ever dry-aged. My dad wasn't too keen on the idea because it requires an upfront investment and a big risk: in those 14 to 120 days of dry-aging, if you don't strike the right balance, a lot of expensive product can be lost. And he was right. When we expanded our Leroy facility, I set aside one room specifically for my dry-aging endeavor. I placed the primals in the refrigerated room at the right temperature and assumed that all I had to do was wait and see the magic happen before my eyes.

From my many years on the job, I knew the 35-degree temperature would naturally dehumidify the room, but what I didn't realize was that when all that fresh meat was placed within those four walls, its high moisture level would greatly increase the humidity in that environment, making it impossible to properly dry-age the product. Rather than the beautiful and seamless result I was hoping for, I ended up with a bunch of rotten meat.

It took me a while to get the hang of it, trial and error and trial and more error. Initially, what I was missing was a dehumidifier potent enough to extract moisture from a cold room, which is harder than doing so from warmer temperatures. I also needed some form of air circulation; this is everything because it helps whisk away the moisture that constantly comes to the beef's surface throughout the dry-aging process. As the moisture leaves the beef and is extracted from the room, the meat pulls inward and retracts from the bone and fat surrounding the cut. This sunken appearance tells me that the water is escaping successfully from the beef. All of these factors, combined and working in perfect harmony, help control the decomposition of the meat and set us on the path to dry-aging success.

Why go to all this trouble? Not just to meet the demand that surged at the turn of this century and continued to grow; it's also about the actual end product. The aging process breaks down the beef's muscle and creates a steak that is ten times more tender and has ten times the flavor of its fresh counterpart.

The longer a cut is dry-aged, the stronger its flavors. In essence, it really comes down to each chef's needs and palate. That's why we dry-age beef from 14 to 120 days, to cover the wide variety of requests we get from our clients. If beef is aged past the 120-day mark, within the right environment, very seldom does the product go sour. But as the extra days go by, the meat will eventually become too dry-aged for about 95 percent of our customers, so it doesn't make sense to push it too far past 120 days.

Once the beef has reached its peak—which could be thirty days for one client and ninety days for another—we take the corresponding primal from the dry-aging room and cut through the extra dark and tough outer layer to access the beautiful bright red piece of meat awaiting within. Then we use the bandsaw to cut through the bone, portion the steaks, and ship them off to restaurant and individual clients.

When you have the largest dry-aging room in the country, cuts like the tomahawk or porterhouse have to be aging and at the ready so that when a customer calls and puts in a last-minute order, we can fulfill it without a hitch. We use this beef for our dry-aged burgers as well, so keeping our supply in check is crucial. This product has become more sought after because the end consumer knows more about their food now than ever before. Many can even taste the difference between fresh and dry-aged beef, and hence order it more when out to eat or cooking up something special at home. It has become so popular that restaurants now pride themselves on their dry-aged programs like they do their wine. And I get it. My dry-aging room is not just my pride and joy, it's like my own personal candy store. When I walk into that cold haven, the sweet smell permeating the air is reminiscent of roasted corn and immediately elicits a smile.

**BY NOW YOU SHOULD KNOW** enough about beef's different grades and quality and cuts to have a better sense of what will suit your taste. Next rule to live by: When you have the chance to buy and eat fresh meat . . . why choose frozen? Unless for some reason fresh beef isn't available in your area, going for the frozen option makes no sense to me. That's why I pride myself on cutting and delivering fresh product to our customers. If rather than home delivery or purchasing through the many other retailers we cater to, you prefer to shop from your local butcher, then it's not just about choosing the right beef for you, it's about knowing how to choose the right butcher.

Why is this important? Because your local retail butcher, the one who only sells direct to consumer rather than to restaurants, isn't inspected by the USDA. That means that they are allowed to conduct their retail cutting and processing without a USDA inspector on the premises (note that butchers at supermarkets aren't regulated either). I'm still baffled by this and do not understand why we aren't all under the same inspection umbrella to keep our consumers safe. If I claim something on a Pat LaFrieda label, I have an inspector on-site asking me for documentation to back up each specific claim. This gives you, the consumer, the peace of mind that what I say on that packaged meat you pick up at your store or order online has been federally verified. Without USDA oversight, a retail butcher can basically say anything they want about the meat they're selling—it's local, it's organic, it's grass-fed, it had a happy life—and if you aren't well informed, you wouldn't know any better.

I'm not saying don't buy meat from your local butchers. All I want is for you to have the right tools to choose a reliable retail butcher so that the meat you purchase is the best it can be.

First off, when you walk up to your butcher's counter, look around. Is the counter and surrounding area clean? Does it look sanitized? Then check the butcher. If their white coat looks like they just did roof work, that's a red flag right there. The butcher's cleanliness when it comes to attire and workplace is a key indication of whether you have just walked into the right or wrong place. If they're sloppy, then it is highly likely that their work is sloppy too, and that equates to a higher health risk. Sanitation and food safety go hand in hand.

Let's say the butcher's coat and station are looking good. Next up is engaging them in conversation. And I'm not referring to everyday banter. Sure, get the greetings and niceties out of the way, and then dive into some real meat talk. Since there are no regulations and claims can't be verified, the goal in this conversation is to figure out if you can trust your butcher. This is subjective—it depends on your needs, your comfort level, and what you're looking to purchase—but here are some simple questions that will point you in the right direction. Where is this beef sourced from? Remember, just because the butcher tells you it's grass-fed, that doesn't mean it's local. Who are the growers? If you prefer to eat domestic product, then you want to make sure the growers are here and not in another country. If you don't care either way, that's fine too, though I can't help but encourage you to support our American farmers. In any case, when butchers are proud of where they source their product, it will be easy for them to answer these questions. If they hesitate or give you the runaround with some vague response, consider that another red flag. Not knowing where their product is coming from means that their purchasing method is likely focused on simply getting what's available or whatever is cheapest on the market, with no real connection to the source.

After approving of the butcher's cleanliness, their ability to converse, and their know-how about where your beef is coming from, then it's your turn to put all your newfound knowledge to work. Look at the color of the meat—it should be a nice bright, vivid red. If you're buying Choice or Prime, check the marbling of the cut; you now know there should be some visible intramuscular fat in there (unless it's a filet mignon, of course). Pay attention to where the beef meets the bone. If you're buying a fresh cut, it should be even and not sunken from the bone or fat. Remember, meat retracts when dry-aged, but it should be leveled when fresh. And finally, if you're going for the marinated meat sitting behind the counter window, just keep in mind it is likely meat that is on the verge of going bad.

Marinating beef extends its life just a little longer. And if that doesn't sell, it's likely that it will end up cooked the next day and sold as cooked marinated beef. This doesn't just pertain to butchers. In retail stores, the food in the cooked sections and to-go counters is usually made from product that didn't sell so as to try to turn a profit before the food goes to waste. It's a common practice. I'm not saying it's dangerous. They're careful. No one wants to use a product that has gone bad because they run the huge risk of being shut down if something were to happen. But it's definitely not as fresh as when you buy it yourself and make it at home. These are all tricks of the trade, like when restaurant-goers ask for a well-done steak—chefs aren't going to waste their prized goods on a well-done order because they know that particular customer won't be able to tell the difference, so they will likely choose to cook up a slightly lower-quality piece instead of the cream of the crop.

Don't be confused by meat labeled "butcher's choice" or "prime rib." When product is not verified, terms are often wildly thrown around to get a customer's attention, but the word *prime* doesn't necessarily refer to USDA-graded Prime beef. Just like with sirloin—some customers often get sirloin confused with strip loin, which is much higher in quality, but they wouldn't know the difference if their butcher sold them a cut of lesser value. That's why it's so important to know your meat.

When shopping at supermarkets, I would choose the prepacked products from federally inspected companies over products packaged at the store without a brand name to back up their quality and level of accountability. Established brand-name products are not only packaged under USDA guidelines, they're backed by companies that are unwilling to take unnecessary risks with what they're selling. One small recall could spell that particular brand's demise. It happened in New Jersey a few years back when several people died due to an unsafe product. After that recall, no one would dare touch that brand and the company went under. That's why brand names matter; there's an accountability factor that is tied to their products and will make those companies tend to err on the side of caution.

At the end of the day, the best butcher shop, supermarket, purveyor, or brand is the one that carries the grade of beef that you're looking for and makes you feel safe and comfortable with their product. I would never have my kids eat meat that is not federally inspected, with butchers wearing dirty coats and using machines that may or may not be cleaned frequently. I would much rather rest assured that my meat is being sourced humanely and processed following precise federally mandated sanitation and safety guidelines.

Now it's up to you. You have all the information you need to choose the best quality and cuts, so flip the page for some inspiration on how to bring the glory of beef to your own kitchen.

# CHAPTER 11

# RECIPES: COOKED TO PERFECTION

After all the beef talk, from how it's processed to where to buy it and how to choose it, the time has finally come to bring that glorious beef to your very own table. Whether you're looking to prep a quick dinner, spend the day grilling outside with your friends, or set up a special night of celebration, these recipes will have you covered. Beef is so versatile you can sear, roast, grill, stir-fry, sous vide it, and so much more. Just remember these two rules to live by:

1. Never pepper your steak before searing or grilling. If cooked at high heat, pepper will cast a slightly bitter flavor to your beef, so add it freshly ground *after* you've served your meat, like the waiters do at fine-dining establishments.
2. Always make sure your grill or pan is smoking hot before cooking for the perfect sear. The biggest mistake when cooking meat is putting it on a surface that's not hot enough. This will steam your beef and make it lose all the flavor it should have.

Now you're ready. I can't wait to see what delicious meals you cook up, so don't forget to tag me @patlafrieda if you decide to post your glorious results on Instagram. Happy cooking!

# SEARED DRY-AGED NEW YORK STRIP STEAK

*If you ask me to choose between a fresh or a dry-aged New York strip steak, I will choose dry-aged all the way. There's no match for that intensity of flavor. And I know I'm not alone. The growing demand for dry-aged beef is what inspired me to custom build the biggest dry-aging room in the country, proof that more people are falling in love with this type of beef. But rest assured that the following recipe works well with either type beef, so don't hold back if all you've got in your fridge is a fresh steak. As for the cut itself, many wonder: Is it better to have a larger filet and less quality in the strip or vice versa? Chefs' opinions vary greatly depending on their personal preferences, but there is really no right answer here. It all comes down to your particular taste.*

**SERVES 2**

2 dry-aged New York strip steaks (12 ounces each)

½ teaspoon kosher salt

2 tablespoons blended oil (75% canola oil, 25% olive oil) or canola oil

2 tablespoons unsalted butter

3 garlic cloves, crushed

1 sprig rosemary

2 sprigs thyme

Freshly ground black pepper

---

1. Pat the steaks dry with paper towels and season each one on both sides with the salt.
2. Heat a cast-iron or stainless-steel pan over medium heat until very hot, just about to smoke, about 5 minutes, then add the blended oil.
3. Place the seasoned strip steaks in the pan and sear for 2 to 3 minutes on each side.

4. Once both sides are seared, add the butter to the side of the pan and allow it to melt until it starts to foam. Add the garlic, rosemary, and thyme and baste the steaks with the foaming butter for about 1 minute on each side.

5. Remove the steaks from the pan and allow to rest on a cutting board or plate for at least 5 minutes, then season with pepper and serve.

**MEAT MASTER TIP:** When ordering this cut from your butcher, ask for a first or center cut, Choice or Prime, bone-in New York strip. I like the bone in a New York strip because although it doesn't add the flavor that some chefs like to imagine it does, it protects the steak from heat. The meat closer to the bone is usually a little rarer because the bone acts as a nice insulator, so for those who like their steaks medium to rare, you'll get an extra boost of juiciness along the bone. As for sides, this perfectly seared steak would go beautifully with a baked potato and sautéed green beans or sautéed spinach with minced garlic.

# GRILLED T-BONE STEAK

*The T-bone steak is basically located right after the strip and before the porterhouse. In the primal that produces these three beautiful steaks, when the filet becomes a little narrower but there's still about a half-dollar diameter present in the cut, it's called a T-bone steak; however, it could easily be served at restaurants as a porterhouse because of their similarities. Nevertheless, a beef connoisseur should know the difference. A T-bone isn't about how much filet it carries; it's really about the great strip quality it delivers. I also enjoy ordering this particular steak because its impressive wide T-shaped bone gives it a nice presence on the plate.*

**SERVES 2**

1 T-bone steak (32 ounces)
2 tablespoons blended oil (75% canola oil, 25% olive oil)
1½ tablespoons kosher salt
Freshly ground black pepper

---

1. Pat the steak dry with paper towels.
2. Preheat a gas or charcoal grill to medium-high heat, 375 to 400°F.
3. Drizzle both sides of the steak with the blended oil and season with the salt.
4. Place the steak on the grill and cook for 2 to 3 minutes on each side for medium-rare. If medium is preferred, add 1 minute of cooking time on each side.

5. Remove from the grill and allow the steak to rest for at least 7 minutes, then season with pepper and serve.

**MEAT MASTER TIP:** Make sure you get a T-bone steak that has at least a half-dollar diameter of filet on it. If they cut too far to the front, there is no filet, and that would make it a New York strip, which is just as good, but costs less. If you're going to pay for a T-bone, make sure you get what you're paying for. USDA Choice or Prime is the way to go for the best flavor, and a side of grilled asparagus would be a wonderful complement to this juicy steak.

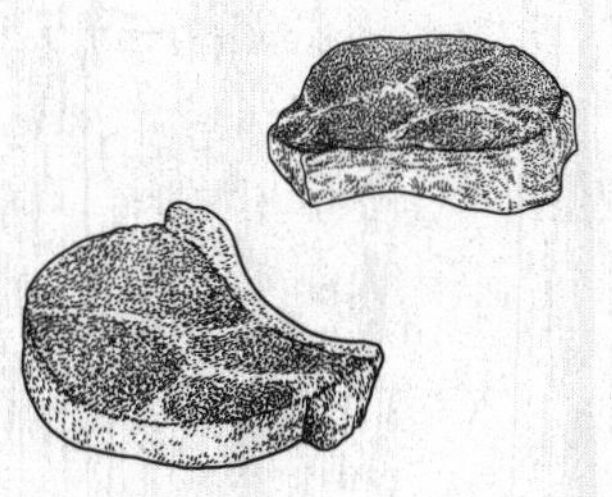

# GRILLED DRY-AGED BURGER

*Brisket is the key ingredient in our original blend of choice. It has a sweet flavor profile and holds up well when chopped to form a burger. However, its fat content is obscene, so it must be mixed with other cuts, such as chuck, to get the perfect burger. Once I started selling Shake Shack their special burger blend, with brisket as the key ingredient, this cut got me into so much trouble in the industry. Up until then, it used to be considered an economy cut used by the barbecue sector, but when we started selling these beef patties to Shake Shack's more than one hundred locations, with 30 percent of their beef coming from brisket—well, we ended up affecting the cattle price structure because of our increased demand. We made it jump from $1.40 to $4.00 a pound. Nowadays, those who want to follow in Shake Shack's footsteps know that brisket is part of their secret, so they use it in their mixes too. We can really blame it on my grandfather, may he rest in peace, because if it hadn't been for his original burger recipe, there would be no famous Shake Shack burger as we know it.*

**SERVES 2**

2 dry-aged burgers (6 ounces each)
1 tablespoon blended oil (75% canola oil, 25% olive oil) or olive oil
½ teaspoon kosher salt
¼ teaspoon freshly ground black pepper
2 slices American cheese (optional)
2 potato buns, split
Lettuce, tomato, and sliced red or yellow onion (optional)

---

1. Preheat a charcoal grill to medium-high heat, 375 to 400°F. If cooking on a stovetop, preheat a cast-iron pan over medium heat. This will give the burgers a great crust and beautiful sear.
2. Drizzle the burgers with the oil. Then season each on both sides with the salt and pepper.

3. Place the burgers on the grill and cook for 3 to 4 minutes on each side to desired doneness. Remove from the grill, set on a cutting board or plate, top each burger with a slice of cheese, and allow to rest for about 3 minutes.

4. Meanwhile, if you enjoy a toasted hamburger bun, place the potato buns facedown on the grill on medium-low heat and toast for 1 to 2 minutes. Place the burgers between the buns, add lettuce, tomato, and sliced red or white onion, if desired, and enjoy.

**MEAT MASTER TIP:** I recommend you come straight to the source and order our own Pat LaFrieda dry-aged burgers made from our special 80/20 blend that will hit you with tremendous levels of flavor on your first bite. However, if you decide to go to your butcher, just make sure the burger blend is coming from real cuts of meat rather than leftover scraps, and ask them to throw in a little chopped brisket to round out your patty with a touch of delicious buttery flavor.

# ROASTED TOP ROUND, SPLIT AND TIED

*In the 1980s and early '90s, roast beef made from the top round was probably the most commonly used meat for sandwiches. It actually has a number of usages. When cut thick, a roasted top round could be considered a London broil. When you reach the center, it's usually raw enough that if you cut a thin slice and add a little sprinkle of salt, you'll get a very similar taste to when my dad used to thinly cut a slice of raw top round and hand it over to me as a treat, essentially feeding me carpaccio, though as a kid I had no idea that was the word for it. Since it has always been fairly inexpensive, top round was also one of the cuts used at the historic beefsteak dinners in the late nineteenth and early twentieth centuries. And, on a personal note, this is the first meat I ever cut at our family's shop, as I described in chapter 1, so it will always have a special place in my heart.*

**SERVES 8 TO 10**

1 top round, split and tied (about 7 pounds)
1½ tablespoons kosher salt
1 tablespoon freshly ground black pepper
2 tablespoons blended oil (75% canola oil, 25% olive oil) or canola oil

---

1. Season the split top round with the salt and pepper.
2. Preheat the oven to 350°F.
3. Heat a large ovenproof pan over medium-high heat until smoking hot, 3 to 4 minutes. Add the blended oil.
4. Place the top round in the pan and sear each side for 3 to 4 minutes, until golden brown.

5. Place the pan in the oven and roast for about 45 minutes, until the internal temperature reaches 110 to 115°F.

6. Remove the pan from the oven and allow the top round to rest for at least 20 minutes before carving.

**MEAT MASTER TIP:** When buying top or inside round, try to choose one that is already tied so you don't have to struggle with that step at home. And make sure to follow the grading. The top round is lean to begin with, so Choice or Prime would be ideal. Anything lower than that and you'll have to slice it very thinly to get a similar effect. There's not much intramuscular fat, so it's a good, healthy cut of meat to eat. Add thin slices of this top round to a fresh baguette and top it with horseradish cream for a sandwich bursting with flavor.

# SOUS VIDE BONELESS BEEF SHORT RIBS

*Boneless beef short ribs are also known as chuck flap tail. One time, I went to eat at Mario Batali's Babbo when I was selling them a lot of bone-in short ribs. To my surprise, when I ordered his brasato al barolo, there was no bone in these beef short ribs on my plate. So, the following day, I asked Mario, "What are you doing with the bones? Do you just toss them?" Turned out they were basically cutting them out. "You know, I could save you about 33 percent food cost if you use chuck flap tail instead, which is the extension of the short rib muscle." And the change was made. It just goes to show how important it is for the butcher to be in sync with restaurant chefs regarding what they're using because we can really help them pick the right cuts for their food costs without sacrificing flavor.*

**SERVES 6**

6 pounds boneless beef short ribs (aka chuck flap tail)
1½ tablespoons kosher salt
1 teaspoon freshly ground black pepper
3 tablespoons blended oil (75% canola oil, 25% olive oil) or canola oil
1 large onion, peeled and chopped
1 medium carrot, peeled and chopped
1 stalk celery, strings removed, chopped
2 tablespoons tomato paste
1 cup red wine
1 cup port wine
1 cup beef stock
1 sprig rosemary
3 sprigs thyme
2 bay leaves

---

1. Preheat a water bath to 170.5°F. Make sure to choose a container with a lid. The beef short ribs will need to cook covered for a full 24 hours.

2. Season the beef short ribs with the salt and pepper. Heat the oil in a Dutch oven over medium heat, add the seasoned short ribs,

and sear for 3 to 4 minutes on each side, until golden brown. Remove from the pot and allow to cool for 15 to 20 minutes.

3. Remove half the oil from the pot and heat the remaining oil over medium-low heat. Add the onion, carrot, and celery and cook the vegetables for about 5 minutes, until browned. Add the tomato paste and cook for about 2 minutes.

4. Deglaze the bottom of the Dutch oven with the red wine and port wine. Reduce the liquid by half, about 10 minutes, then add the beef stock, followed by the rosemary, thyme, and bay leaves. Reduce by half, 7 to 10 minutes, then set aside and allow the vegetable sauce to cool, about 15 minutes. The consistency should be on the thicker side, not too soupy.

5. Once all the ingredients have been cooled, place the seared beef short ribs into 2 gallon zip-top bags together with their resting juices. Add the vegetable sauce to each bag and remove as much air from the bags as possible. If you need help, slowly submerge the bags into the sous vide water container to help push out the extra air. Then seal the bags completely and double-check that there are no leaks.

6. Submerge the bags in the container with water, cover with the lid, and sous vide for 24 hours. The short ribs should be very tender. Serve right away or freeze the bags for later use. To reheat frozen short ribs, defrost, then microwave for 3 to 5 minutes inside the zip-top bag and serve.

**MEAT MASTER TIP:** If you ask for chuck flap tail instead of boneless beef short ribs (which is essentially the same thing in my industry), your butcher will likely be impressed you know the inside terminology. No need for an expensive vacuum sealer for this recipe—large gallon zip-top bags will work just as well. These ribs go spectacularly well with homemade creamy mashed potatoes, rice pilaf, or celery root puree.

# SPICY HANGER STEAK PIZZAIOLA[1]

*My grandmother introduced me to the world of long hot peppers. When I would visit, she'd sauté some long hots and give them to me as a snack with a fresh loaf of sesame knotted twist Italian bread before whatever meal she was cooking up that day. For those of you who don't usually eat spicy food, these peppers will be slightly hot, but if you're used to having your meal with a little kick, long hots are totally bearable and absolutely addictive. I couldn't get enough of them. Some nights my grandmother would make hanger steaks (which my grandfather liked a lot, although not as much as his beloved skirt steak) and add some sautéed long hots, and I would be in dinner heaven. I'd also go to bed happy knowing that if there were any leftover peppers, she'd surely add them to the following morning's omelet, which I loved because it added a touch of unexpected heat to that breakfast staple. This is my friend and renowned chef Rocco DiSpirito's take on my childhood classic.*

**SERVES 4**

4 hanger steaks (6 ounces each), trimmed
Kosher salt
Freshly ground black pepper
1 tablespoon extra-virgin olive oil
8 garlic cloves, thinly sliced
1 small onion, sliced ½ inch thick
1 tablespoon chopped fresh oregano
2 long hot red peppers, sliced ½ inch thick
1 cubanelle pepper (a long slender pepper), sliced ½ inch thick
1 cup chopped tomatoes (if you use processed tomatoes, choose a brand with no fat, sodium, or added sugar)
¼ teaspoon crushed red pepper flakes (optional)

---

1. Preheat the oven to 350°F. Place a wire rack over a baking sheet.
2. Pat dry the surface of the steaks with paper towels and season with salt and pepper.

3. Pour the olive oil into a large ovenproof skillet that has a lid, place over high heat, and heat until it smokes. Add the steaks and brown each side evenly, about 2 minutes per side. Remove the steaks and let rest on the wire rack. Reduce the heat to medium.

4. Add the garlic to the pan and cook, stirring, until browned, about 2 minutes. Add the onion, oregano, and hot and cubanelle peppers and cook until softened, about 3 minutes. Add the tomatoes and cook until they form a sauce, 4 to 5 minutes.

5. Place the steaks on top of the tomato-and-pepper mixture, cover, and place in the oven. Cook for about 5 minutes for medium-rare to medium. Remove the meat from the sauce and allow to rest on the rack. On the stovetop, cook the tomato sauce down until it's thick, about 8 minutes, then stir in the red pepper flakes, if desired. Cut each of the steaks into three chunks and divide among four plates. Spoon the sauce over the meat and serve.

**MEAT MASTER TIP:** When ordering hanger steak from your butcher or buying it at the market, make sure to avoid the ungraded cuts and instead go straight for the one graded Prime or Choice. It's not a very tender piece of meat, so if you go ungraded or buy a low-quality cut, it will really be a tough experience, pun intended!

# EPILOGUE

## A TWIST OF FATE: PLANS, A PANDEMIC, AND NEW OPPORTUNITIES

I don't think my dad ever envisioned our company growing like it did. He just couldn't see it, but I could, thanks to him. Even forty years ago, when we were a small company with a barely recognizable name in the industry, driving around town in a couple of beat-up vans with the "Eat My Meat" slogan that my dad had drummed up in the 1970s, those who did know who we were knew that we were trustworthy—a reputation my dad had built through his honesty, loyalty, charisma, and approachability. To this day, our word is our bond. This solid foundation gave me the confidence to hit the streets in search of new clients, it inspired me to take measured chances, to invest in the equipment that would help us cater to our clients' needs, and to innovate along with our industry so that we can continue to keep up with whatever demands come our way. I was able to turn the potential I saw as a young man into a reality and showed my dad that the sky really was the limit. As I've shared with you in these pages, it wasn't an easy journey. Convincing my dad to take chances and believe in change was an uphill battle most of the time, but when he saw that what I was doing was working, he readily embraced our progress. So much so that now my dad likes to say:

"I spent all of my money on your education for this reason. You were destined to be in this business!"

"No, Dad, it's because Aunt Lisa asked you and forced you to bring me into the business," I remind him, jokingly. We always have a good laugh with our back-and-forths.

My dad could have retired long ago, but he continues to come to work every morning at 3:00 a.m. and stays until about noon. After his first cup of coffee while watching the news at full volume so his near-deaf ears can make out the hot topic of the day, he goes over invoices, fixes stuff around the facility, oversees what we're all doing, and never shies away from speaking up to voice any opinions, no matter how un-PC they may be . . . and everyone loves him for it. I swear he and my cousin Mark can get away with saying anything and still remain in everyone's good graces. I think my dad, like his father, will likely spend his last years happily standing next to a butcher table at work. It's in our blood.

I knew what we were capable of but am still astounded when I take a second and let it all sink in. We are forty times larger than when I joined in 1994, growing to a point of servicing 1,500 restaurants, as far west as Las Vegas, as far south as Miami, and even crossing the border into Mexico. And we've earned it. We put in the work six days a week, twenty-four hours a day. My clients know that they can reach out to me directly any time day or night and I will always take their call and will be there for them. If they're in a bind, I will drop everything and help coordinate whatever delivery they need to come through for them. I actually love that adrenaline rush, that feeling of making the impossible happen.

That's also why I love when we're called to work an event like a music or dance festival that draws thousands of people. It's tireless work—my team and I hardly pick our heads up when we're manning one of those stands—but the thought that our speed is one of the factors that will determine if people will choose to give our stand a shot is thrilling. You know how it goes—when you're at a festival and starving, if the pizza line is too long, you and your buddies are going to take a look around and see if there's any spot with a shorter wait or one that at least seems to be moving along quickly. I get it, standing in line frustrates me, so the quicker we can service our customers the better. And the satisfaction of feeding thousands of people in one go is priceless.

I love that personal connection with our customers—it provides me with the tangibility and contact that I so yearned for during my months on Wall Street. That's why I also love it when clients drop by our facility for a tour. Nothing like pulling back the curtain to show them everything that goes into running the business that processes the high-quality meat that they receive neatly packaged, labeled, and boxed up in their kitchens. They inspire me to go above and beyond, and I hope our product inspires them to create unforgettable dishes for their patrons.

Our steady yearly growth and enduring goal to continue to supply protein to present and future generations pushed us to think even bigger and dive into the construction of an additional facility two blocks north of our original one, which will allow us to triple our production overnight. But it wouldn't have been possible without our consistent principle of always putting our money back into the business. Rather than pocketing our profits and buying yachts or other lavish merchandise, we've paid our bills, kept our debt down, saved for rainy days, and constantly reinvested in our company. We don't shy away from purchasing new equipment to better streamline our production process, new trucks to serve our rising client list, and building a new facility to accommodate our growth and hopefully feed more than the three hundred thousand people we already reach daily. That kind of expansion takes time. And that dedication, passion, relentless work ethic, and business mentality of reinvestment and diversification is exactly what saved us in March 2020 when the world suddenly came to a standstill.

When COVID-19 hit, we were smack in the middle of the construction of our new $20 million building, had just opened our third location at the Time Out Market in Brooklyn, were planning to service a dance festival down in Miami, and had endless projects and possibilities on the table. Then came Friday, March 13, 2020. New York was on lockdown. The Pennsy and our three restaurants in the Time Out Market were suddenly shut down. We were left scrambling, making sure the food in our fridges didn't go to waste and following all the necessary protocols to properly close up our locations until further notice. The collective state of shock was palpable. By Monday, March 16, we had lost $32 million. The shutdown meant that my long roster of restaurant clients had to lock up their venues too, overnight, without a clear end in sight. In the following days, the calls poured in from restaurant owners who explained that they had to stop payment on the checks they had recently mailed to us, which were intended to cover their outstanding invoices. Everyone was bracing themselves, quietly observing how the situation evolved. Some chefs stashed their freezers with meat so that they could be ready when the lockdown was lifted. I'd get calls that went like this:

"Pat, I need a favor, these are the dimensions of my freezer. I need you to send me whatever you can to fill it all the way to the top."

"Okay, sure. But what's the favor part? Whenever you order meat, don't I always give it to you?"

"Yeah, but don't you have a break in supply?"

"No, but we do have corona prices," I said, messing with him in those early days when being lighthearted still seemed feasible.

None of us knew then how badly this pandemic was about to affect the restaurant industry. That's why many chefs opted to pack up their belongings and take off to their summer homes to ride out the viral wave at a distance, thinking it would only be a matter of weeks . . . not months. Meanwhile, left with a $32 million financial burden practically overnight, I sprang into action and was relieved I had already stocked up on Purell knowing that if the city remained paralyzed, we would still be called to work through those days in order to continue supplying food. I figured the industry would be back up and running by April or May at the latest.

Nevertheless, Pat LaFrieda Meat Purveyors remained open and we marched forward. When everyone retreated to their homes, my team and I were on the front lines making sure the meat supply chain continued to flow smoothly. We remained on our regular six-days-a-week schedule, I installed hand sanitizer dispenser units in every entrance to every room, and I took all necessary safety measures as they were rolled out to ensure my employees were safe. Yet there was one thing I had no control over . . . as the weeks rolled by and the restaurant industry remained on lockdown, I had no choice but to reduce our staff to get through the peak of this emergency. It took years to build our army of 180 employees, and my big high in life is to create jobs, not to take them away, so to see 75 of our soldiers suddenly gone was devastating. I knew it was a temporary sacrifice that had to be made so that we'd be ready to bounce back even faster as soon as we received the anxiously awaited green light to do so, but that didn't make it any easier. There were moments where I truly felt heartbroken. My staff has been amazing throughout this crisis, and I couldn't have done it all without them.

With our roster of restaurants all but disappearing, we suddenly shifted to what had up until then been a smaller part of our business: retail and direct-to-consumer sales. If there was ever a lesson to be learned in that moment it is that diversification is absolutely essential. It's one of the key factors that allowed us to survive those trying months. When the 60 percent of New Yorkers who usually fed themselves by dining out suddenly veered toward supermarkets for their main food source, both big and small retailers were suddenly overwhelmed by the massive influx of customers emptying their shelves on a daily basis. These retailers in turn needed to find new sources of food to keep their shelves stocked. Thankfully, we were able to adapt quickly, having already ventured into retail a few years earlier. While our restaurant clients remained in limbo leaving past-due invoices hovering over my head like a cumbersome dark cloud, I amped up our retail sector and suddenly 98 percent of our business was coming from LaFrieda home delivery, Amazon, ShopRite, and Fresh Direct. We already had the packaging machines in place, but as orders increased, I took down a wall in our facility and installed an extra machine just to keep up with the demand. This unexpected yet welcome turn of events is what allowed us to remain afloat as the weeks turned into months. The thing with retail is that although the orders may be huge, they also require extra time because, contrary to wholesale requests from restaurants, each portion must be individually wrapped and labeled. We were working with reduced manpower, but I rallied my troops and we made it happen, managing to consistently fill each and every order that came in.

Nevertheless, fear began to spread that there would be a meat shortage, and I was suddenly flooded with questions about the possibility of a break in our supply. I got a call from the head of Shake Shack's food chain supply in early March.

"Pat, I need to know what your contingency plan is if there's a break in supply."

"Listen, we could lose half of our staff and still be operational and you'd still get your product. I'd worry about your customers and employees."

I spent hours consistently reassuring clients, the media, and friends not to worry. The world may have been on lockdown, but our meat supply chain would not be broken by this pandemic. I don't know how many times I have had to repeat this: there was never a meat shortage. Did the meat supply chain break down? No. The Tyson and Smithfields plants that closed also reopened, some only after a few days. Furthermore, one closed voluntarily to avoid bad press and the other had no choice, but it was only one of the company's facilities out of many. No food company wants to be seen as carrying and distributing unsafe product, so those two closings led to some other voluntary and temporary closures, which in turn led to a reevaluation of operational protocols and the eventual reopening of these facilities at 50 percent capacity. However, this doesn't mean there was less meat to go around. The growers still had all their animals, and as soon as the few harvesting facilities that had closed were up and running again, that small bump in the road was overcome, and meat continued to flow to the end consumers. By May it was harder to score toilet paper than chopped beef.

Meanwhile, new projects surfaced with the few restaurants that were up for the fight of their lives to survive, such as Mohawk House in New Jersey. The owner and I were brainstorming on the phone when restaurants were closed down, and I suggested he be the bridge between our meat and his customers. I sent him a truck with a small amount of product to start with, resupplied consistently, and we advertised it on social media. We also agreed that he would upcharge accordingly, but there would be absolutely no price gouging because that's not who we are. When I broached this subject, I didn't mean to insult him in any way because I know he's a standup guy, and I apologized if it had come across in the wrong way. But since my name was going to be associated with this venture, I simply wanted to make absolutely sure we were on the same page.

I got so much amazing feedback about our truck and what it did to reduce food stress and panic in that community, and it was such a success that Shake Shack called me soon after wondering what they could do along similar lines. And that's when I suggested the burger kit, which included the raw burger patties, buns, sauce, and cheese. They loved the idea, and we jumped right in. We initially put a cap of 207 kits per day, and they sold out in fifteen minutes. So they called me and asked me to double the cap, and the next day we sold 414 kits. By April, our cap had increased to 700, which was the most we could put out in one day. Keep in mind that we were the ones buying the bread and cheese, taking five-pound bags of sauce and reducing them to three-ounce packets, and putting these ingredients plus our patties into each individual kit, then shipping them through Goldbelly. It was time-consuming, but worth it. We were moving product, helping Shake Shack, and delivering a little ray of joy during the darkest days of the outbreak.

Then the Shake Shack bailout debacle happened, and overnight we went from 700 to 230 kits a day. It was like a switch went off for them. Even though they returned the $10 million bailout that was originally intended for small businesses, they had to navigate some hard press and had a tough few weeks ahead of them. They weren't the only ones in that predicament, but they were unfortunately the first ones to get their name in the press wheel. Thankfully, they survived.

As if a pandemic weren't enough, in early April, our company got hit with a virus of its own, in the computer system, which suddenly erased LaFrieda from the books. Everything was gone. It's a miracle I didn't have a heart attack that weekend. I couldn't believe my eyes. Then I found out that an exchange of emails had begun between the IT company that managed my server and the extortionists. This wasn't a simple glitch or a case of someone downloading something they shouldn't have by mistake. These were actual extortionists who had targeted my company and were now demanding I pay up $400,000 in three days to get my system back up and running. If I didn't come through, the ransom would be doubled. I couldn't believe it. First thing I did when I hung up with the IT representative was contact the FBI, the North Bergen Police Department, and New Jersey's Cyber Crime Unit. I would be damned if in the middle of a pandemic I gave my hard-earned cash to a group of extortionists. I needed it to continue operating my business. Did they not know what was happening with the restaurant industry and how that was affecting my company? I had no idea what was going to happen next. I felt so vulnerable. After a few hours of absolute uncertainty, the IT company said they thought there was a backup server they might be able to salvage. This was a Saturday. It took the entire weekend to deal with this crisis, and the weight of it combined with the effects of the worldwide pandemic was almost unbearable. We weren't operational again until Monday, but I got my data back without paying those opportunistic criminals.

In the meantime, as we approached May, there were high hopes that the New York lockdown on restaurants in particular would be lifted, so my clients began to take care of some of their past-due invoices to clear the way for a possible reopening. Things were looking slightly up until the lockdown was extended into June. That was a blow for us all. Small businesses usually survive paycheck to paycheck—they barely make it on a good day, so how were they expected to make it now and how long would it take the industry to actually recover? These questions are still up in the air as I write this.

Worry began to take root in my mind. If restaurants didn't make an eventual comeback, I thought I might not make it out alive either. But retail continued to skyrocket, and since we barely had any trucks on the street, our numbers were somehow balancing out. By July, my wrists hurt from cutting so much meat. Amid the extended uncertainty, we continued to be busy, which was a good sign. Restaurants were still lagging behind, with only outdoor dining as an option now. Some were able to make it work, especially because they thought indoor dining would soon follow, while others began to drop out of the race.

The next unexpected blow came when I found out that Time Out Market had to shut two of my three restaurants down permanently to repurpose the space given the extenuating circumstances. It was another heartbreaker. I was so proud of those spots and all we had accomplished there, and now we were back down to one. Thankfully we managed to reopen it in August, and it became busier than it ever had been pre-COVID, so all was not lost. Come September, home delivery and retail continued on a strong streak—who would have thought the one sector my dad and I didn't cater to as fervently in the past was now the one saving our asses? Shake Shack had also bounced back, and we added a smoked burger kit to the menu, which included bacon and hot peppers—another hit. Over the summer, I had also joined forces with Source Brewing for a special limited-edition Fourth of July beer, with proceeds benefiting the Intrepid Museum, and it sold out before it was even available to the public. It was such a runaway success that we decided to collaborate on a special fall edition, called the Butcher's Brew, and I'm looking into the possibility of distributing Source Brewing's product with our fleet of still mostly parked trucks, which would help me bring more of my employees back to work too. I'm always open to new ideas and collaborations, and this one has proven to be a wonderful surprise.

We're not out of the woods yet. The initial $32 million we lost in past-due invoices has now come down to around $26 million. It's slow progress, but it's something. And since restaurants were allowed to have indoor dining at 25 percent capacity starting in October 2020, some of them were able to attack their debt with us a little further. What will happen next remains to be seen, as that small percentage may still not be enough for many to justify the operational costs of remaining open.

Despite this difficult year, I continue to look ahead. After overcoming what felt like endless pandemic setbacks and slowdowns, our new building is finally done and operational, which means that soon I will have my most prized possession up and running:

our new state-of-the-art dry-aging room. I'm talking a 125-foot-long by 18-foot-tall by 18-foot-wide room exclusively dedicated to the magic of dry-aging. To put this into perspective, our original dry-aging room, which was already considered the biggest in the country, was able to hold around 5,000 primals, weighing on average twenty-five pounds each, and our new one will be able to hold 12,500.

I had been dreaming about this expansion for years, so when the time came to go for it, I decided to design this particular room myself—I knew exactly what I needed for it to work and flow as seamlessly as possible. First, I built it to scale in my office, and my father looked at me like I was on drugs. But I kept hammering out each detail to carefully maximize the space, and now it's a reality. When I walk into that room and hop in the vehicle that transports me all the way to the back, turns, and comes back to the front, I revel in having been able to fulfill my vision of efficiency. I'm also thrilled with the two brand-new desiccant cooling systems, which are basically units that encompass a complete system of ventilation, temperature control, and dehumidification to keep our product stable, at the right temperature, and dry. And they're set up outside (which I love because that keeps them from taking up valuable space inside), each with dual twenty-inch-diameter ducts—one absorbs air with moisture from the room and the other delivers air without moisture to keep the ideal balance for drying beef.

Since we started dry-aging meat twenty-plus years ago, demand has consistently increased. It felt like we could never get the beef ready fast enough. We filled our original room to capacity, which only served to highlight the need for a bigger and better space. The new building had been calculated to be finalized just in time to accommodate our growth. However, when COVID-19 appeared, it was like slamming on the brakes. The few restaurants that did remain open for takeout and delivery weren't interested in much dry-aged beef because it's the type of meat that is meant to be enjoyed at the table, a luxury item in the food world. We definitely still move some product, but not at the rate that we did pre-COVID. It will likely take a while for it to rebound. And in the meantime, we have to be careful with our demand calculations because it really is a big investment of time and money.

When I order beef, I have to pay for it within seven days from the time it leaves the harvesting facility's loading dock, and, as I've mentioned in earlier chapters, sometimes it can take up to four days to get to me. By the time the primals for dry-aging are placed on the shelf, I've already paid for that product. From then on, I'm taking a risk, banking on the restaurants to come through with their orders 30, 60, 90, and 120 days later. Such a huge investment makes me carefully crunch the numbers and keep an eye on the demand so that at the end of the cycle, I'll be able to actually sell the product. It's trickier to project now given our present circumstances. So, although we have already transferred our product to the new room and phased out the old one, it will likely take longer than originally forecast to fill that monumental space to capacity.

As with any major crisis, recovery takes time. I don't doubt the restaurant industry in New York City will eventually make a comeback, but it won't happen overnight. I think we may even see a boom of high-end restaurants in the suburbs, turning unexpected locations into new food hubs.

It's not all rosy days in the meat industry. It takes long hours and hard work. Each and every day we have to prove ourselves to our clients, and sometimes we don't get it right the first time around and we have to go back to the drawing board because our customers stake their reputation on us, and our reputation is tied to their experience. Failure is not an option. That's why we deliver six days a week. That's why there's no order cutoff time. We cut and pack each order with the utmost care, whether it's going to a restaurant or a family home. We pride ourselves on our high quality and the ability to keep that superior level consistent throughout the years. No one can take that away from us. Are we perfect? No. Do we make mistakes? Yes. Do we fight for our customers? Most definitely.

Despite all the obstacles I have faced, I have never thought of giving this business up. Fear of failing, maybe. The idea of giving up, never. From the first day I walked into our shop, I knew that was going to be my life. And I was excited about it. Wall Street was more like an experience than a job or a career because my heart has always been at LaFrieda. I was once offered a job on the floor of the stock exchange itself. It was a big job. I shadowed the guy for two days. All the screaming—it was nuts in there. I just couldn't picture myself doing that and being happy. I ran into this same guy around five years later at the bar of a restaurant that was one of our customers. He saw me and said, "Pat LaFrieda. You must be kicking yourself in the ass for not taking that job, huh?" And I calmly replied, "No, I'm doing fine. I'm happy." We had already increased our business, but we weren't by any means where we are today. Yet I was doing what I loved most, and nothing can replace that feeling. The arrogance of that man's assumption only motivated me to grow more.

The pandemic actually made our retail sector excel to levels I never dreamed of, but I will never be able to celebrate this moment given the tragedy and loss we've experienced this year. So, what's next? What now?

In the short run, we will continue to focus on growing our retail clients and catering to them. I can't wait for restaurants to fully rebound; they are an essential part of our existence, but for the time being, I estimate we will continue to have less demand for product in New York City and more demand in the suburbs and other cities. Does that mean New York is dead? Far from it. The city is going to suffer, but it will rise to the challenge and come back stronger. And when the restaurant industry begins to thrive again, we will be ready and waiting to welcome them back.

In our many years of consistent growth, we've spearheaded the burger craze, expanded into retail and restaurants of our own, distributed our product beyond the tristate area, created our own signature steak sandwich and sauce, and even starred in our own Food Network reality show, *Meat Men*. The power of a family-run business is that we don't have to answer to a board and there's no red tape—there are risks, but they're ours for the taking. We've come an incredibly long way and, at the same time, it feels like the tip of the iceberg. When panic strikes, we come up with new solutions that open up unexpected doors and take our business down unforeseen paths of innovation and prosperity. I thrive on the thrill of these emerging possibilities.

But at the end of the day, I'm just a butcher, following my grandfather's invaluable mantra: *Don't deviate from what you do, just do it better every day.* That's what has gotten us this far, and that's what will continue to inspire us as we move forward to whatever comes next. Butchering is not my job, it's my life.

*Pat LaFrieda,*

NOVEMBER 2020

# ACKNOWLEDGMENTS

First and foremost, thank you, Dad. None of my accomplishments would have been possible without the wisdom and passion you shared with me for business and life. All I ever wished for was to make you proud. It is within this lifelong pursuit that I fell in love with bridging the gaps between people and meat. Thank you for raising me to be the man I am today. I owe you everything.

If it weren't for Dan Halpern, Ecco's founder and longtime publisher, I wouldn't have even contemplated the idea of writing this book. Thank you, Dan, for your unflinching motivation and guidance, and for helping me give life to *Glorious Beef.*

To Johanna Castillo, my editor turned literary agent, thank you for once again blowing me away with your tenacity and passion to ensure the seed of this idea became a published book. I'm fortunate to have you and your relentless drive in my corner.

Thank you, Cecilia Molinari, for listening to my stories and occasional rants, night after night, at LaFrieda's facility, and for turning them and the patchwork of behind-the-scenes industry descriptions into a cohesive narrative. You listened, consulted, and balanced a complicated enigma. It would've never happened without you.

To Brandon Bryan, my executive chef at Pat LaFrieda Meat Purveyors, thank you for stepping up to deliver the recipes featured in this book with your perpetual can-do attitude, even during the New York restaurant industry's most challenging times. Your loyalty will never be forgotten. Most of all, thank you for your friendship.

Thank you to my team at Ecco, for your tireless work and enthusiasm, especially Gabriella Doob, my editor, for your careful, precise, and thoughtful edits, and for seeing these pages through to the finish line.

And last but far from least, to my team at LaFrieda: Thank you for standing steadfast by my side through the hardest moments imaginable. Your support, loyalty, and willingness to go the extra mile embody the spirit of Pat LaFrieda Meat Purveyors.

# NOTES

## 1: The Growers


1. "Beef Charts," USDA Agricultural Marketing Service, accessed October 7, 2020, https://www.ams.usda.gov/reports/meat-percentage-charts.
2. "Organic Livestock Requirements," USDA, accessed October 7, 2020, https://www.ams.usda.gov/sites/default/files/media/Organic%20Livestock%20Requirements.pdf.
3. "Becoming a Certified Operation," USDA Agricultural Marketing Service, accessed October 7, 2020, https://www.ams.usda.gov/services/organic-certification/becoming-certified.
4. "Farm Resources, Income, and Expenses," USDA National Agricultural Statistics Service, accessed May 24, 2021, https://www.nass.usda.gov/Publications/Todays_Reports/reports/fnlo0220.pdf.
5. "2017 Census of Agriculture," USDA National Agricultural Statistics Service, accessed May 24, 2021, https://www.nass.usda.gov/Publications/AgCensus/2017/Full_Report/Volume_1,_Chapter_1_US/usv1.pdf.
6. "Farm Resources, Income, and Expenses," USDA National Agricultural Statistics Service.

7. "Land Values 2019 Summary," USDA National Agricultural Statistics Service, accessed October 7, 2020, https://www.nass.usda.gov/Publications/Todays_Reports/reports/land0819.pdf.

## 2: The Feed Effect

1. "The Ruminant Digestive System," University of Minnesota Extension, accessed October 7, 2020, https://extension.umn.edu/dairy-nutrition/ruminant-digestive-system#stomach-compartments-1000460.
2. "Ruminant Nutritionist," AGcareers.com, accessed October 7, 2020, https://www.agcareers.com/career-profiles/ruminant-nutritionist.cfm.
3. "Food Safety and Inspection Service Labeling Guideline on Documentation Needed to Substantiate Animal Raising Claims for Label Submissions," USDA Food Safety and Inspection Service, December 2019, https://www.fsis.usda.gov/sites/default/files/import/RaisingClaims.pdf.
4. Michelle Miller, "Farm Babe: Should Cattle Eat Corn? Myth vs. Reality in Livestock Nutrition," *AG Daily,* November 29, 2016, https://www.agdaily.com/livestock/farm-babe-cattle-eat-corn-myth-vs-reality-livestock-nutrition.
5. "Acidosis," Beef Cattle Research Council, accessed May 24, 2021, http://www.beefresearch.ca/research-topic.cfm/acidosis-63.
6. "Animal Feeding Operations," USDA Natural Resources Conservation Service, accessed October 7, 2020, https://www.nrcs.usda.gov/wps/portal/nrcs/main/national/plantsanimals/livestock/afo.
7. Dave Weaber and Mike Miller, "An Evolving Industry," *Beef,* September 1, 2004, https://www.beefmagazine.com/mag/beef_evolving_industry.
8. "Overview of Greenhouse Gases," US Environmental Protection

Agency, accessed October 7, 2020, https://www.epa.gov/ghgemisions/overview-greenhouse-gases.

9. "Cows vs. Cars?," Dairy Cares, YouTube, August 28, 2019, https://www.youtube.com/watch?v=RW8BclS27aI.

10. "The Biogenic Carbon Cycle and Cattle," UC Davis CLEAR Center, February 19, 2020, https://clear.ucdavis.edu/explainers/biogenic-carbon-cycle-and-cattle.

11. "Livestock's Long Shadow: Environmental Issues and Options," Livestock, Environment and Development (LEAD) Initiative, UN Food and Agricultural Organization, 2006, p. 272, accessed October 7, 2020, http://www.fao.org/3/a0701e/a0701e.pdf.

12. "Sources of Greenhouse Gas Emissions," US Environmental Protection Agency, accessed October 7, 2020, https://www.epa.gov/ghgemissions/sources-greenhouse-gas-emissions.

13. "Time Line of the American Bison," U.S. Fish and Wildlife Service National Bison Range Wildlife Refuge Complex, accessed October 7, 2020, https://www.fws.gov/uploadedFiles/Bison%20Fact%20Sheet.pdf.

14. "Why Regenerative Agriculture," Regeneration International, accessed October 7, 2020, https://regenerationinternational.org/why-regenerative-agriculture.

15. "Nationwide Shift to Grass-Fed Beef Requires Larger Cattle Population," IOP Science, July 25, 2018, https://iopscience.iop.org/aticle/10.1088/1748-9326/aad401.

16. Nancy Matsumoto, "Is Grass-Fed Beef Really Better for the Planet? Here's the Science," NPR.org, August 13, 2019, https://www.npr.org/sections/thesalt/2019/08/13/746576239/is-grass-fed-beef-really-better-for-the-planet-heres-the-science.

17. Amy Quinton, "Cows and Climate Change: Making Cattle More Sustainable," UC Davis Feeding a Growing Population, June 27, 2019, https://www.ucdavis.edu/food/news/making-cattle-more-sustainable.

18. "Kansas Ranchers Dispute UN Report That Links Cows to Climate Change," CBS News, August 8, 2019, https://www.cbsnews.com/news/un-climate-report-ranchers-dispute-report-that-links-cows-to-climate-change.

19. Angus Chen, "Gassy Cows Warm the Planet. Scientists Think They Know How to Squelch Those Belches," NPR.org, September 22, 2017, https://www.npr.org/sections/the-salt/2017/09/22/552698446/gassy-cows-warm-the-planet-scientists-think-they-know-how-to-squelch-those-belch.

## 3: Harvest Time

1. Dave Roos, "The Juicy History of Humans Eating Meat," Dave Roos, History Channel, June 20, 2019, https://www.history.com/news/why-humans-eat-meat.

2. "Humane Methods of Slaughter Act," USDA Natural Agricultural Library, accessed October 7, 2020, https://www.nal.usda.gov/awic/humane-methods-slaughter-act.

3. Temple Grandin, "Design of Cattle Stun Boxes," YouTube, August 3, 2008, https://www.youtube.com/watch?v=mCYs1CSsbqs.

4. "Creekstone Farms," Ozersky.TV, accessed October 7, 2020, Vimeo, https://vimeo.com/46883305.

5. "An Introduction to Mobile Slaughter Units," USDA Food Safety and Inspection Service, February 21, 2017, https://www.usda.gov/media/blog/2010/08/30/introduction-mobile-slaughter-units.

6. Gene Kim and Jessica Orwig, "Only 60% of a Cow Is Actually Used for Food—Here's the Weird Stuff That Happens to the Other 40%," *Business Insider,* October 29, 2017, https://www.businessinsider.com/surprising-everyday-products-made-from-cow-parts-2017–10.

7. "Fun Facts: Products We Get from Beef Cattle," Beef2Live, ac-

cessed September 5, 2020, https://beef2live.com/story-fun-facts-products-beef-cattle-0-104636.

## 4: The Importance of Grading, Labeling, and Traceability

1. "United States Standards for Grades of Carcass Beef," USDA Agricultural Marketing Service, December 18, 2017, https://www.ams.usda.gov/sites/default/files/media/CarcassBeefStandard.pdf.
2. "Code of Federal Regulations," govinfo.gov, January 1, 2014, https://www.govinfo.gov/content/pkg/CFR-2014-title9-vol2/xml/CFR-2014-title9-vol2-sec310-22.xml.
3. "United States Standards for Grades of Carcass Beef," December 18, 2017.
4. "High Line History," The High Line, accessed October 7, 2020, https://www.thehighline.org/history/#:~:text=The%20first%20train%20ran%20on,meat%2C%20dairy%2C%20and%20produce.
5. "Cattle and Beef: Sector at a Glance," USDA Economic Research Service, accessed August 4, 2020, https://www.ers.usda.gov/topics/animal-products/cattle-beef/sector-at-a-glance.

## 9: What's the Beef with Eating Meat?

1. Dave Roos, "The Juicy History of Humans Eating Meat," History Channel, June 20, 2019, https://www.history.com/news/why-humans-eat-meat.
2. Jeffrey Kluger, "Sorry Vegans: Here's How Meat-Eating Made Us Human," *Time*, March 9, 2016, https://time.com/4252373/meat-eating-veganism-evolution.
3. Paulette Gaynor, "How US FDA's GRAS Notification Program Works," *Food Safety Magazine*, US Food and Drug Administration, accessed May 24, 2021, https://www.fda.gov/food/

generally-recognized-safe-gras/how-us-fdas-gras-notification-program-works.

4. "Poverty: Overview," World Bank, accessed April 16, 2020, https://www.worldbank.org/en/topic/poverty/overview#:~:text=There%20has%20been%20marked%20progress,less%20than%20%241.90%20a%20day.

5. "Decent Rural Employment: Livestock," UN Food and Agricultural Organization, accessed October 7, 2020, http://www.fao.org/rural-employment/agricultural-sub-sectors/livestock/en.

6. "Growing at a Slower Pace, World Population Is Expected to Reach 9.7 Billion in 2050 and Could Peak at Nearly 11 Billion Around 2100," UN Department of Economic and Social Affairs, June 17, 2019, https://www.un.org/development/desa/en/news/population/world-population-prospects-2019.html.

## 11: Recipes: Cooked to Perfection

1. Recipe by Rocco DiSpirito. Copyright © 2020 by Flavorworks, Inc. All rights reserved.